Packaged Composite
Applications

Packaged Composite Applications

Dan Woods

O'REILLY®

Beijing · Cambridge · Farnham · Köln · Paris · Sebastopol · Taipei · Tokyo

Packaged Composite Applications
by Dan Woods

Published by O'Reilly & Associates, Inc., 1005 Gravenstein Highway North, Sebastopol, CA 95472.

O'Reilly & Associates books may be purchased for educational, business, or sales promotional use. On-line editions are also available for most titles (*safari.oreilly.com*). For more information, contact our cor-porate/institutional sales department: (800) 998-9938 or *corporate@oreilly.com*.

Editor:	Dale Dougherty
Production Editor:	Darren Kelly
Cover Designer:	Edie Freedman
Interior Designer:	David Futato

Printing History:

June 2003:	First Edition.

ISBN: 0-596-00552-0
[M]

Contents

Introduction

As an executive of SAP, an international firm, I have the privilege of traveling around the world and talking to people at every level in every sort of company. I have three kinds of conversations. Some business executives tell me about strategy without mentioning how it will be implemented. Some CIOs tell me about innovative technology without mentioning its business value. The conversations I enjoy the most are about both topics: how strategy can be realized with technology or how technology creates new capabilities that can improve a company's competitive and strategic position.

This book tells the story of how management and technologists can work together on both a strategic and tactical level to chart a course toward a better infrastructure. While the current economic environment and the history of enterprise computing dictate that evolution will occur in small increments, one step at a time, the most important factor for senior management, CIOs, and technologists is their vision for the future.

SAP's vision is that a new layer created on top of the current infrastructure will transform the heterogeneous mélange of legacy systems and products from multiple vendors into a unified environment on which a new generation of applications can be built. This vision produced a technology blueprint: the Enterprise Services Architecture. The unifying layer is called an ESA platform.

The new applications are Packaged Composite Applications (PCAs), software that snaps onto current systems and enables automation to expand across enterprise application silos. At their heart, Packaged Composite Applications are simply about getting the most value out of current systems at the least possible cost. To do this intelligently requires reducing the cost of adjusting and optimizing business processes to meet competitive challenges by building flexibility into the infrastructure.

The story is told in context. The reader is guided through how PCAs fit into the history of enterprise computing, how the organization will change as cross-functional automation becomes affordable, how vendors will react, specific business scenarios in which PCAs make a difference, and how PCAs and the ESA platform will evolve. Of course, the detailed structure of PCAs is examined.

This book is unique not only in the vision it sets forth but also in how it was constructed. We gave the author, Dan Woods, unprecedented access to the top technical and business minds in SAP and asked him to create an authoritative description of the idea of Packaged Composite Applications. We asked O'Reilly & Associates, the most respected publisher of technology books in the United States and Europe, to apply its standards of clarity, precision, and editorial quality to the production of this book. The result is a lucid presentation of how this new paradigm for creating applications will likely change the face of corporate IT.

PCAs are not an SAP product. They are a style of creating an application, a style that is inspired by the current realities facing the enterprise. Many other vendors will be offering PCAs and their own version of the ESA components. The SAP branded products are SAP xApps and SAP NetWeaver. The importance of PCAs and the Enterprise Services Architecture vision is not that it meets the needs of a particular vendor, but that it meets the needs of customers. We chose to present this material with an independent author and publisher so that the reader would be confident that all sides of the concept would be examined. You won't find SAP mentioned much at all in this book. What you will find is a well-crafted analysis of what we believe may be the best way for companies to use IT to win in the marketplace. We encourage analysis and debate on this topic and are eager to hear your views.

This book breaks new ground for SAP. One major implication of PCAs and ESA is that developers will be invited into a deeper level of the SAP infrastructure and into the ongoing conversation about where that architecture is going. Saying that you are open to communication is one thing. Acting that way is what persuades people. This book on PCAs, a second title on ESA later this year, and a series of initiatives for the developer community in 2003 and 2004 will show that SAP is changing. It is becoming more open with its ideas and its architecture, encouraging different kinds of partnerships with ISVs, systems integrators, and technology vendors, and bringing more companies into the process of making SAP software work for its customers. We hope that you will join us in this effort.

—Shai Agassi
Member, SAP AG Executive Board

Preface

This book is the result of a systematic search through the brain trust of SAP for all of the relevant arguments, examples, concepts, and analogies related to Packaged Composite Applications. As an author, it was a delight to have access to authoritative sources at every level: from the vision of PCAs for the next 10 years, to the requirements engineering process, to the technical architecture, right down to developers who struggle to knit everything together. The ideas in this book belong to these talented men and women and others outside of SAP who enthusiastically participated in the mission of this book: to explain PCAs and their likely impact on the world of Information Technology.

The book also benefited substantially from interviews with leading analysts like Josh Greenbaum of Enterprise Applications Consulting and Yvonne Genovese of Gartner Research. Another valuable perspective came from executives at systems integrators including Roger Ford and Shannon Tassin from Accenture, William Grasham from Deloitte Consulting, and Adolf S. Allesch from Cap Gemini Ernst & Young.

The chapter on the history of the IT infrastructure was brought to life with the experience of James D. Robinson, III, General Partner of RRE Ventures; Nancy Martin, partner at Warburg Pincus; Dr. Ravi Kalakota, CEO of E-Business Strategies; and Alan Cooper, author of *The Inmates Are Running the Asylum*.

The collaborative process between O'Reilly & Associates and SAP produced a book that is unprecedented in its scope and vision. The joint publication arrangement allowed O'Reilly's exacting editorial viewpoint inside the planning process of SAP. This allowed us to reach beyond the traditional forms of technology communication. This book is not a marketing treatise about neatly shaped colored boxes. It is not a backward-looking, outdated description of a product without context. Rather, this book combines the approach of a forward-looking analyst with the perspective of an executive who must make things work, without skimping on the relevant technical details. We examine the ideas driving PCAs forward in the marketplace and the problems and solutions that an executive and technologist will encounter in implementation. The result, both SAP and O'Reilly sincerely hope, is an authoritative text that allows all interested parties to assess the value of PCAs for their lives as executives, technologists, analysts, sales representatives, and users.

Acronyms Used in This Book

For your reference, here is a list of the acronyms used in this book.:

CRM	Customer Relationship Management
EAI	Enterprise Application Integration
ERP	Enterprise Resource Planning
ESA	Enterprise Services Architecture
HR	Human Resources
ISV	Independent software vendor
PCA	Packaged Composite Application
PLM	Product Lifecycle Management
ROI	Return on investment
SCM	Supply Chain Management
TCO	Total cost of ownership

How to Contact Us

We present strong arguments in favor of PCAs in this book, but you may want to join in the debate. We welcome your voice. You can contact us at:

O'Reilly & Associates, Inc.
1005 Gravenstein Highway North
Sebastopol, CA 95472
(800) 998-9938 (in the United States or Canada)
(707) 829-0515 (international/local)
(707) 829-0104 (fax)

To ask technical questions or comment on the book, send email to:

bookquestions@oreilly.com

We have a web site for the book, where we'll list errata and any plans for future editions. You can access this page at:

http://www.oreilly.com/catalog/pkgcompaps

For more information about this book and others, see the O'Reilly web site:

http://www.oreilly.com

Acknowledgments

This book is dedicated to the men and women of SAP, whose ingenious ideas and bold vision made researching and writing this book an exquisite pleasure. Shai Agassi, Peter Graf, and Tim Bussiek, the devoted sponsors of this project, helped it along at each stage with comments, encouragement, and suggestions on how to find the right information in the vast brain trust of SAP. Pascal Brosset, Mark Sochan, Dennis Moore, Kevin Fliess, Rochelle Frey, and Ranjan Das all went beyond the call of duty. They diverted themselves from jammed schedules, understood the mission of book, and kept it on course with detailed points of view. Sami Muneer was this book's best friend; without him, the project would have failed. In addition to his full-time job, he arranged an endless parade of meetings and interviews, reviewed every page of the text, and improved the book substantially through his comments. I hope he writes a book soon because I would love to read it. I would also like to thank Dale Dougherty and Mark Jacobson of O'Reilly for bringing me into this project, and Deb Cameron, whose advice and coaching helped teach this author some new tricks.

—Dan Woods

1

The PCA Paradigm

This book is about Packaged Composite Applications, a new architectural paradigm for enterprise applications. In the following pages, we will tell the story of PCAs: what they are, how they work, and what they may mean for executives and technologists who strive to extract business value from technology.

Most new technology ideas are exciting because they offer new possibilities. PCAs are different. The argument of this book is that while PCAs offer compelling potential for innovation and for expanding the support for key business processes, there is another factor that is more important: PCAs are inevitable as a next step for IT.

Many will scoff at this assertion, or perhaps dismiss it as undue enthusiasm for a promising idea. But as we will explain, a variety of technical and business forces are converging to produce this inevitability.

The wake of the dot-com boom, the gradual maturing of the IT infrastructure through Enterprise Resource Planning (ERP), the rise of best-of-breed enterprise applications such as Customer Relationship Management (CRM) and Supply Chain Management (SCM), and competitive pressure on companies to improve their performance all combine to lead to the development of PCAs.

Briefly summarized, the argument for PCAs is as follows:

- There is no starting over. Existing technology, which was lovingly maintained and improved for Y2K, must be leveraged, not discarded, in the evolution to the next level of productivity.

- Infrastructure investment will be driven primarily by applications with a demonstrable return-on-investment (ROI) and will not take place for its own sake without compelling cost savings.

- Funding for technology is under pressure and will remain so. Any way that total cost of ownership (TCO) can be reduced will be welcomed.

- Due to the high costs of maintenance and a long history of failure to meet expectations, custom development will have a high bar for approval.

- Optimizing processes within the silos of enterprise applications or between them using hard-to-change integration technology has reached its practical limits. The low-hanging fruit is gone. The next generation of applications must flexibly automate processes across all application boundaries, easily and affordably. Doing so will reduce the cost of change and increase the pace of innovation.

- The next frontier for competitive advantage is automation of strategic and decision-making processes beyond the capabilities of best-of-breed enterprise applications. This sort of automation requires flexible cross-functional automation and must also bridge the gap between transactional and collaborative applications.

As we will explain in detail in the following pages, these arguments imply a world ripe for the adoption of the PCA paradigm. The implication is that new applications must cost less, produce ROI quickly, leverage existing technology, automate cross-functional processes, bridge collaborative and transactional systems, and have the TCO of packaged software rather than the higher maintenance of custom development.

PCAs Defined

The most formal definition of PCAs is that they are the application delivery layer of the Enterprise Services Architecture (ESA). ESA is a blueprint for how applications can be created on top of existing enterprise applications to increase the value of those systems to the enterprise, to extend automation to new processes, and to enhance the ability of corporations to innovate. The ESA vision is implemented in an ESA platform, which is shown in relation to PCAs in Figure 1-1.

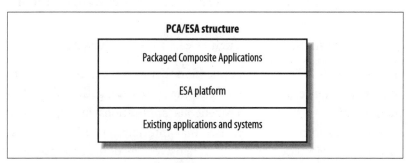

Figure 1-1. Enterprise Services Architecture

Note that Enterprise Services Architecture takes in all three of these layers, while the ESA platform represents the software that sits in the middle layer of this architecture, interfacing PCAs with existing enterprise applications.

This diagram provides only a gross overview and we'll drill down in considerable depth as we move along. For now, let's examine PCAs themselves in more detail. From the bottom up, we can break down PCAs as follows:

"Applications" means:

- Performing new application functionality beyond simply integrating existing systems.
- Expanding support to flexible, configurable, collaborative processes beyond the scope of existing applications.
- Allowing analysis followed by action, not read-only aggregation of corporate information.

"Composite" means:

- Aggregating functionality of existing systems, exposed through web services.

- Crossing traditional application boundaries.
- Creating a comprehensive process and information model.

"Packaged" means:

- Products with a lower TCO than custom development.
- Products with integration costs leveraged over a large customer base.
- Products supported with new releases and maintenance.

In other words, a PCA is an application that uses the ESA platform to unlock the value of a company's existing systems. It gathers the information from all of the heterogeneous applications into a unified, homogenous form, and then uses that information to build a new, focused solution based on a comprehensive view of the enterprise.

The benefits of this approach are easy to see, and we will explore them in detail in Chapter 6. A project management PCA, for example, reaches into the human resources (HR) system to find out the expertise of staff available for assignment, collects budget information from the financial systems, retrieves detailed project schedules from end-user project management applications like Microsoft Project, and assembles a master view of all corporate projects. This master portfolio shows the performance of projects across many dimensions, beyond what is held in any single project management repository. The collaborative functionality allows a larger group of people across departments to be brought into the process, and the rich information keeps them there. Structured as well as unstructured information is managed and linked to points in the process so that an institutional memory is created about when and why decisions are made. As a result, the unrealized value of the underlying HR, financial, and project management applications is unlocked. The information and functionality of these applications have become services to the PCA and can be used by a much broader subset of the enterprise, leveraging the investment in existing technology. Other examples of PCAs include systems for product definition, mergers and acquisitions, and employee productivity.

But how does this happen? On the surface, it is easy to see how PCAs would be grand and wonderful. But veterans of the IT battles of the past 20 years know that in the real world, there is not one HR system, but many, probably from different vendors, probably in different states of repair. The same is true for finance, supply chain, customer relationship management, and ERP. Only a rookie would assume an easy solution for the problems inherent in a heterogeneous mix of vendor applications and legacy systems.

Enterprise Services Architecture

The ESA vision is the next step forward in a long march of progress that has its roots in the idea of a service-oriented architecture. It is enabled by the improving capabilities of business process management software and will be brought to life with the help of web services, portals, and Enterprise Application Integration (EAI) technology.

The job of the Enterprise Services Architecture is to make a collection of heterogeneous systems look homogenous so a PCA can do its job. An ESA platform, a product that implements the ESA vision, assembles all of the technology just mentioned and provides a two-way street between a PCA and the underlying enterprise applications. To do this, the ESA and PCA must be able to model the enterprise, create unified repositories of information gathered from heterogeneous systems, generate workflows and business processes on top of this unified information that reaches into the underlying applications, and employ the functionality of existing systems as services. Of course, all of this must be presented to the user in a coherent and simple user interface.

Whew! One might think that proving the unified field theory or building the Great Wall of China would be good practice before facing a task like creating an ESA platform and then building a PCA on top of it. But remember that evolution works in mysterious and wonderful ways. Parts of an organism evolve separately, developing new capabilities until all at once—bang!—something happens to coalesce all of the developing functionality in a revolutionary advance.

A variety of forces are conspiring to transform ESA from an attractive idea to deployable technology. We have the rise of the Internet, which proves the power of simple technology like HTML and HTTP and creates the world's largest service-oriented architecture with web servers, domain name servers, and a host of supporting services all working together in a massively decentralized but finely coordinated way. HTML leads to the creation of XML, which becomes a pervasive presence, the lingua franca of IT. Standards for business process management and other sorts of orchestration are gradually taking form. XML standards emerge within vertical industries as well, helped along by organizations like RosettaNet, trade associations, and B2B exchanges. A variety of other forces lead vendors and companies of all sizes and shapes to create inter-application communication techniques based on XML, to create business process management tools that separate process logic about business issues from application logic about technical details, and to find ways to model and unify data. Then—bang!—web services appear, create a standard method for describing how a service works, and bring everything together.

The biggest impact of web services is not the technology, which is still in its early stages of development, but the change in attitude it has brought about. Now the entire IT world believes that all applications can be wrapped in a service and talk to each other. EAI vendors must find this grating because their approach to XML messages can justifiably claim much of the same territory as web services, with a variety of advantages. But EAI did not cause people to believe in a new world of ubiquitous services. Web services did.

The job of vendors who create ESA platforms becomes much less daunting with all of the coordinated effort. With every vendor creating software that offers or consumes web services or both, the set of new functions required for ESA becomes smaller and more manageable. We'll all build the Great Wall together, one section at a time. The problem shrinks further when you consider how much existing technology for portals, data unification, workflow, and business process management can be leveraged to create an ESA platform.

Who will provide this platform? The usual suspects, of course—the large companies we all know and love will combine their existing technology into ESA offerings. Technology vendors like IBM, Microsoft, and Oracle, suite vendors like SAP and PeopleSoft, independent software vendors like Seibel and Documentum, and systems integrators like Accenture and CapGemini will all get into the act, as we discuss in Chapter 5.

What this ESA layer will look like in detail will vary from vendor to vendor, but at a macro level the ESA will allow a PCA to:

- Build a user interface for the application using standard components.
- Create a unified repository of data collected from the heterogeneous underlying systems.
- Create new repositories for information collected by the PCA.
- Store relationships between data in the underlying systems.
- Expose underlying systems as web services to make functionality and structured information from those systems available to the PCA.
- Store unstructured information.
- Create new workflows and business processes using all these elements.
- Allow collaboration across all of these elements.

There is much to be said about how all this will be accomplished, and the details of implementation and the approach to the ESA platform are crucial to making the right plans. The task of explaining and analyzing ESA is so large that we are devoting a forthcoming book to that topic. In this book, we look at ESA from the vantage point of the PCA developer.

Given the trend toward incremental improvement, rather than replacement, of existing technology and the desire of companies to invest in small applications that provide a quick ROI, we believe that vendors will not try to sell the ESA platform as a product initially. Instead, individual PCAs that solve specific points of pain and can be implemented quickly will be the leading edge of this revolution. These early PCAs will come packaged with the needed ESA functionality until ESA platforms are released as development platforms in their own right.

The argument of this book is that the most relevant path forward for most companies will be to look at business problems through the lens of PCAs. Companies don't want to talk about platforms. They want to talk about making money.

Flexible Cross-Functional Automation

We will spend a lot of time in this book discussing automation. Chapter 2 describes the history of automation and the demands the enterprise will make on the infrastructure. Chapter 3 looks at arguments for and against delivering more automation through PCAs. Chapter 4 explains exactly how the automation delivered by a PCA is constructed. It details how PCAs are built on top of the ESA platform. In Chapter 5 we discuss how PCAs will change the enterprise and how the technology vendors will handle their emergence. Chapters 6 and 7 are devoted to describing business scenarios and some initial PCA offerings. Chapter 8 evaluates the future of PCAs.

To properly understand how PCAs work and how they fit into the current IT infrastructure, we will take a closer look at the difference between the rigid cross-functional automation present in the current architecture and the flexible approach enabled by PCAs.

In the current IT environment, automation has been extended from the core financial and control applications at the center of the ERP platform to a variety of enterprise applications, including CRM, SCM, HR, Product Lifecycle Management (PLM), and the like. Each of these enterprise applications automates a business process within a silo, a specific domain of functionality. When it comes time to automate a process like order to cash, the end-to-end process may require that several of these applications (CRM, SCM, and ERP, for example) participate. An order may be taken in CRM, fulfilled in SCM, and billed through the ERP system. The transfer of information between the enterprise applications is accomplished through hard-wiring these applications together using integration technology that is expensive to create and difficult to modify. The result is cross-functional automation, but at the cost of rigidity, as illustrated in Figure 1-2. A detailed explanation of why this approach is rigid can be found in Chapter 2.

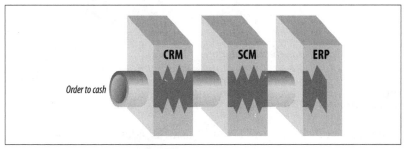

Figure 1-2. Rigid cross-functional process

PCAs change this picture by replacing hard-wired integration with flexible process automation. A PCA, using the ESA platform, automates processes using functionality from business process management, application servers, and a variety of other application component systems that are amenable to change and configuration. The enterprise applications play a role in these automation processes, but the hard-wiring is avoided. As a result, it is easy to automate, modify, and improve processes that range across all enterprise applications. PCAs, therefore, enable configurable processes that span application boundaries: flexible, cross-functional automation (see Figure 1-3).

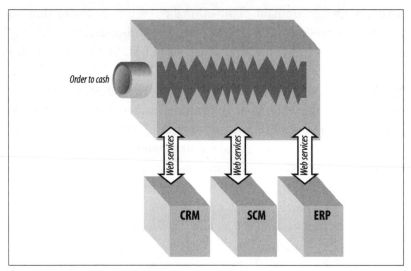

Figure 1-3. Flexible cross-functional process

The Business Case for PCAs

The marriage between the IT and business sides of most companies is frequently stormy, with each partner wary of the other. There is plenty of evidence of failure and bad behavior, but divorce is not possible. Technology and business must find a way to work together to solve an organization's problems. When this happens elegantly, with trust, empathy, and creativity on both sides, amazing things like Wal Mart's supply chain or Amazon's shopping experience emerge and provide massive business value.

PCAs offer a dynamic new force to align business and technology because they offer a way out of the tradeoffs of the past. The situation faced by most companies is a political standoff inside the corporate fortress with the competitive barbarians at the gate.

The Current IT Standoff

The nature of the current infrastructure at many companies is that either each department in a company has its own set of applications, or centrally provided applications represent a lowest common denominator of functionality.

The problem with each department owning its own applications is that the same functions are replicated across departments. The company ends up paying for licenses for 20 ERP implementations when 5 would do, because departments cannot agree on a common implementation.

When applications are centrally provided, departments generally desire more customization and make up for the lack of functionality by extending the central applications, frequently under the radar through a hidden budget line, so the department gets exactly what it wants.

Admittedly, we've oversimplified the problem, but we believe that this example gets to the essence of why IT managers are frequently unhappy. They are either being yelled at by the CEO because there is so much duplication, or the departments are whining because they are not getting exactly what they want. Any dramatic change faces stiff opposition, creating a standoff. Departments resist centralization to avoid losing functionality while the corporate office advocates it to reduce costs.

Looking outside the company to the competitive environment, the situation is tough all over for most businesses. The downturn in the economy has squeezed budgets at every level for IT projects. After five years of furious funding and the successful resolution of the Y2K bug, IT is now being told to make do with what it has. Revenues have dropped and staff has been reduced, but competition has not abated. Customers are in the driver's seat and they know it.

Cost pressure has repeatedly trimmed staff, systems, and suppliers until there is no obvious place to cut. Outsourcing can provide relief but may require a costly retooling of systems. Applications like CRM have been implemented and helped squeeze costs out of the system, but the first squeeze of the lemon always produces the most juice.

The next wave of improvement for most businesses will come in two areas: extending automation to strategic processes and decreasing the cost of change. Progress toward both of these goals is enabled through flexible automation of cross-functional business processes. Businesses will be challenged to evolve faster and to play a stronger game in coordinating activity across departments to better serve customers and cut costs. But the current systems are not up to the job and custom development projects are too expensive to contemplate.

PCAs Offer a Way Out

PCAs break the standoff and empower the company to fight the barbarians. They offer a way forward by extending the current infrastructure using packaged software without disrupting existing applications.

With the ESA platform unifying a heterogeneous collection of applications, new functionality can be created using a PCA. PCAs are products built to be configured from a flexible set of services. PCAs craft a solution out of the tools provided by the ESA platform. Instead of green field development, changing functionality of PCAs is a matter of adding business content such as taxonomies that describe the information being manipulated, and then configuring the PCA, with some additional consulting perhaps in exposing existing systems as services.

When faced with a new business challenge, a flexible infrastructure can react without lengthy delay or huge costs. Businesses can start innovating again. The bar for support of new ideas is lower. The time to market is shorter, which means more ideas can be implemented and more will have a positive ROI. It also means that the existing infrastructure is leveraged to provide more information and functionality to more users.

The collaborative emphasis of PCAs and their ability to unify information across existing systems allows PCAs to support strategic processes that require a comprehensive view of the enterprise. This support avoids time-consuming assembly and rollup of information from component systems, records the decisions for later steps in the process, and manages group communication. All of this contributes to strategic efficiency, the essence of which is making decisions as fast as possible based on the right information.

PCAs also allow organizational consciousness to spread across departmental boundaries. While many versions of a specific enterprise application like HR, ERP, or CRM may exist in a company, each class of application is a world unto itself, separate from other classes. CRM and HR both have employee information, but the functions each application performs and the information stored is often completely distinct. PCAs spin a web across classes of applications and create a comprehensive model of information and functionality from each. These cross-functional processes can be the key to larger efficiencies or competitive advantages.

A PCA for Product Definition

The product definition process provides fertile ground for illustrating the benefits of PCAs. Product definition at most firms begins with the collection of ideas and customer requirements for new product concepts or improvements on existing products, and continues with an evaluation of the concepts and a definition of the form they would take in the marketplace. The next step is a detailed market and technical analysis for each concept, whether for new products or for product improvements. Product concepts that are approved move into prototype design and development, followed by a market testing phase, and finally into production.

In the best case, developing new products is risky. On average, one out of four products brought to market becomes a commercial success. While competition and unforeseeable market forces contribute to unsuccessful product introductions, many mistakes can be avoided. The process is ripe for automation by a PCA. Given the huge expense involved in product launch, improving the product definition process so that more products succeed would represent a tremendous benefit.

Research into product definition shows that the most common product definition mistake is insufficient market analysis. Successful launches spend almost twice as much time and resources on analyzing the market for a potential product than do failed ones. Collection of ideas tends to be ad hoc and undervalues sources for ideas such as employees or suppliers. When ideas are submitted, there is rarely follow-up or reward programs that encourage submission of more ideas. Once gathered, a small team, often just one person, evaluates the ideas and requirements. The evaluation team frequently lacks the interdisciplinary skills and the deep technical knowledge to find all of the show-stopping problems early in the process.

The second largest mistake in product definition is failing to uncover significant technical or design issues soon enough so that development resources are not wasted. In large companies, a commitment to develop a given product is often made before product definition efforts are adequately visible. This lack of visibility early in the process means that efforts may be duplicated and that relevant research is not shared. Resources for research and analysis are diluted by product concepts that are not promising or not aligned with strategy. Rollup and categorization of portfolios of ideas across divisions may never occur because of the expense of combining inconsistent data formats, which means interdependencies or contradictions between product concepts go unnoticed. Other problem areas include timing of product release and managing too many new products simultaneously.

When companies do implement heavyweight processes for product evaluation, they are frequently rigid, cumbersome, and require more work than is needed for simple evaluations. It is not unusual for such processes to be side-stepped, resulting in pet projects being fed into the product development cycle without undergoing an appropriate level of evaluation. Resource allocation may be inconsistent or even capricious.

A product definition PCA could remedy many of these problems. The PCA can standardize the information repository and the processes for product definition across divisions of the company, to bring more people into the process at the right time, and to provide a comprehensive view of the portfolio of ideas, concepts, and products under development.

A product definition PCA would start by creating a standard repository for the ideas submitted from various sources and the information collected from existing enterprise applications. Information about customer requirements would be harvested from the CRM systems and information about specifications for current products would come from the Product Lifecycle Management (PLM) system. Data about market size would be standardized, and unstructured information like research reports would be associated with related ideas. To remedy the problems with certain heavyweight processes, the PCA could implement evaluation processes tailored to different types of ideas—completely new products on the one hand and minor improvements to existing products on the other—and enable collaboration with larger, multidisciplinary teams earlier in the process.

The ideal product definition PCA would transcend a well-organized repository for product concepts. It would create a system for cross-functional collaboration in market and technical analysis not only from R&D, the traditional owners of this process, but also from marketing, legal, environmental, quality management, manufacturing, and anyone else with the skills and expertise to help.

The PCA could be integrated with existing systems to allow information about products, budgets, and marketing schedules to be automatically assembled and extracted. Standard methods of categorization can be applied to ideas across different cycles and the information assembled to evaluate products and potential markets can be preserved for use by future teams looking at similar ideas. Such categorization can help avoid information overload by removing unrelated information from consideration.

The result of such a PCA would be the streamlining and standardization of a vital but ad hoc and frequently inconsistent process. The standardized repository allows information to flow in and out of existing systems so that information collected in the product definition phase can automatically appear in the PLM system to support development. During prototype development, significant findings can be fed back to the product definition repository so that analysis, assumptions, and conclusions about the product and potential market can be reconsidered. This makes product evaluation and development a continuing two-way process, which increases the quality of the market analysis.

Perhaps most significantly, a product definition PCA allows the entire portfolio of ideas under development to be viewed as a whole, ensuring that ideas proceeding through the pipeline are aligned with corporate strategy and market trends. This PCA is representative of the power of this technology: it automates a decision-making, collaborative process, using information from existing systems and providing a comprehensive view where none existed before.

Who Will Build PCAs and Who Will Buy Them?

The P is the letter in PCA we have discussed the least, but it is vital to the value proposition of this paradigm. The P refers to Packaged, meaning packaged as a product, which may seem odd given the flexibility stressed so far. PCAs are easy to productize because of their architectural flexibility, but composite applications need not be products. They can also be created in a consulting or custom development project. ESA platform vendors will gladly sell their software to allow custom composite applications to be built by whoever has the money to pay. Such is the software business.

But the economics of the software business—the productization pendulum we discuss in Chapter 2—will inevitably drive software vendors toward offering composite applications as products, just as every new area of software has been productized.

The preference for software products as opposed to custom development seems only to increase. A March 2003 report from Forrester Research, "Making Packaged and Custom Apps Coexist," points out that 80 percent of CIOs interviewed think that custom applications block intercompany collaboration and 91 percent see custom applications as more expensive than products, mostly because of the difficulty of integration.

The difference for composite applications built on the ESA platform is that products will appear before custom applications, contrary to the normal technology cycle in which custom development is productized. As we explain in Chapter 4, the ESA platform will develop in stages, and in the first stage, vendors will be using the ESA platform for development to the exclusion of corporate IT developers.

For everyone involved in implementing a PCA—from the vendor who develops it to the systems integrator who may configure it to the internal staff who may customize it—there is less programming and more configuration. Every layer that PCAs use is configurable and there is a strict separation between business or process logic, which controls the way the PCA serves the user's needs, and the logic for the services, which contains general reusable functionality as well as state-change and persistence mechanics. Configuration reduces the need for expensive custom coding by allowing the behavior of an application to be controlled by configurable processes, settings, and rules. The result of this is that PCAs are friendly to productization in version 1.0 in a way that most software is in its seventh or eighth release.

The reason for this configurability is the packaged integration and flexibility of the ESA platform. PCAs build on a well-organized set of objects, services, and processes that are built to be flexible. With such a foundation, it is simple to extend that flexibility to the end user.

What productization means for PCAs is exactly what it means for other software products: lower total cost of ownership, professionally maintained software with upgrades and bug fixes, and integration with existing software infrastructure amortized over the entire installed base, including documentation and support.

Who will create PCAs? The entire software industry. Internal development staffs and systems integrators like Accenture, Cap Gemini, and BearingPoint at the most customized, vertical end of the spectrum will use the PCA paradigm without the P to solve business problems quickly. Systems integrators, under pressure to offer smaller projects now that Great Wall-sized efforts are impossible to sell, will put the P back in and make products out of PCAs that will appeal to a broad audience. Independent software vendors (ISVs) will break their programs apart and present them as services for use by PCAs and retool their user interfaces to create PCA-based vertical applications from their technology. Suite vendors like SAP, People-Soft, and Oracle will expose their enterprise applications as service applications that can be used by PCAs. They will create vertical applications after turning their enterprise applications into component services. The suite vendors and the pure technology companies like IBM, BEA, and Sun Microsystems that sell application servers will sell the ESA platform to all of the above. In Chapter 5, we delve into the details of how all of these players will react the arrival of the PCA paradigm.

Who will buy PCAs? The usual suspects from the world of enterprise IT. For all the reasons stated so far—faster ROI, smaller projects, cross-functional optimization, architectural flexibility, lower TCO—corporations will seek the benefits of PCAs once they become convinced they can obtain a strategic or cost advantage. The result for companies who adopt PCAs will be a more flexible deployment model in which the CIO becomes a service broker, selecting among internal services and web services provided by vendors to provision the firm's business needs. The primary task for IT staff will change from API-based programming to semantic modeling connecting web services plumbing. The result should be strategic efficiency and the ability to make decisions and adapt the IT architecture faster and cheaper than previously possible. We explore the impact on the company in detail in Chapter 5.

For everyone, vendors and buyers alike, PCAs represent a lowering of the execution barrier. Because there will be a shorter and less costly trip from thought to fulfillment, the true strategic advantage will come from the vision of how to best serve the customer, the quality of products, and the quality of service. The good news is that ideas will matter more. The bad news is that the competition will catch up faster.

Exactly what forces will affect the competitive environment and what companies must do to win or lose is the topic of the next chapter. To get to the bottom of many of the issues we have just discussed and to see how PCAs can really make a difference, we ask several experts for their views about where IT infrastructure has come from, where it is going, and what companies should do about it.

2

Governing Forces in Enterprise IT

So far, this book has been long on excitement about PCAs. You may have the impression that the author is taken with PCAs much as a cub programmer becomes dizzy with the thrilling potential of a new technology and ignores the practical realities of creating business value.

No indeed. We will not be short on ideas, arguments, and evidence to support our view of PCAs. This book aims not at cheerleading, but at methodical argumentation based on logic and examples. That is why in Chapter 3 we will cover arguments for and against PCAs and discuss potential barriers to wide-scale adoption.

But to make our case effectively, we first must clear the decks and present our view of the major trends in the IT marketplace and how they affect the emergence of PCAs. Using ideas from well-known figures in the technology industry, such as design guru Alan Cooper, former American Express Chairman James Robinson, Nancy Martin, a partner at the private-equity firm of Warburg Pincus, and Dr. Ravi Kalakota, author and CEO of E-Business Strategies, we will examine what the enterprise needs from IT from the point of view of the consumer, the CEO, the investor, and the analyst, and examine the options IT has in responding to those needs. Our goal is to determine whether PCAs can deliver what the enterprise needs today. At the end of this chapter, we explore other approaches to providing the functionality that business requires.

To start off, however, we will survey the evolution of IT infrastructure over the last 30 years.

The Evolution of IT Infrastructure

Over the past 30 years, every generation of IT infrastructure has made progress in four areas. More processes are automated, more functionality is productized, more interfaces are standardized, and more elements are encapsulated in an abstraction.

The first of these areas represents the IT infrastructure's increasing automation of business processes. The last three of these qualities are different aspects of the commoditization of technology, in which, like the formation of sedimentary rock, elements collect together, and relationships between them solidify, gradually forming them into a solid layer that becomes a commodity. The absorption of more functionality into the operating system is a prime example of this trend.

The increase in process automation is easy to understand. The first mainframe applications were focused on financial controls. Now the financial control applications have ballooned into huge ERP systems, and we have product areas to automate almost every part of every business: Customer Relationship Management (CRM), Supply Chain Management (SCM), Human Resources (HR), and on and on. All of these applications extend automation to more business processes.

The increase in productization, standardization, and abstraction is clear but more complex to describe. Typically, a business need like contact management is identified. People write their own little applications to perform needed functions. Vendors then create products for contact management to replace those custom products. Like a one-way pendulum, successful custom applications almost inevitably end up as products, driven by the economics of specialization, the benefit of spreading development costs over the widest possible customer base, and the lower cost to the customer of owning a software product that is maintained and upgraded rather than expensive custom code for which one customer must bear the entire cost.

Standardization is a similar process. A functional need appears, like the need to access data in a database. Many products are introduced to address the need, but the users of databases and software developers are frustrated because they have to learn a new way to use a database for each product. To simplify things, a standard may then emerge, as structured query language (SQL) did to provide a general way to access any relational database. Vendors start supporting that standard and fight about how to extend it.

The increase in abstraction is the gradual separation of technology into more and more layers that build on each other and have standard ways of interacting. In the beginning, programmers stored records in files. Database programs encapsulated that functionality and abstracted away the complications of keeping track of all of the details of which file contained a particular record and where that record was located in the file. The world became simpler for programmers, but database programs grew in complexity. SQL made the interface simpler still, and now Open Database Connectivity (ODBC) and Java Database Connectivity (JBDC) handle more of the details for the programmer. Cathedrals of abstractions now exist in the form of application servers and messaging layers, all of which have become fiendishly complex under the covers but have made life easier for their users.

Commoditization follows productization or standardization. Clones come on the scene when a product has no way to protect itself either through patents or through some other barrier to replication. A successful standard may result in many products supporting the standard to the same level of quality. In all cases, commoditization represents a drop in the amount of money earned by vendors. If commoditization can be avoided but productization and standardization achieved in a large market, companies like Microsoft are the result. Commoditization in the technology business is like entropy, an unstoppable force that consumes all successful technology. More and more functionality is absorbed into sedimentary rock with each generation of progress.

In order to better understand how the next wave of growth in IT infrastructure will come upon us, we will look at the three main stages in the evolution of IT: the mainframe era, the client/server era, and the Internet era.

Mainframe Era: Central Command and Control

The mainframe era, roughly 1970 to 1985, was characterized by a corporate emphasis on central command and control. Financial applications were most frequently automated; a dumb terminal, most often an IBM 3270, connected to a monolithic application that hard-wired everything together (see Figure 2-1). Layers such as the MVS operating system, CICS, and IMS managed various parts of the user interface, application services, and the database, but the layering of the application was primarily up to the application architect, not a matter of principle.

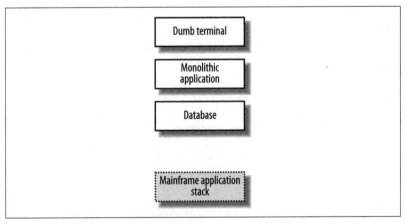

Figure 2-1. Mainframe application stack

The IBM 360 architecture and the standard operating systems built on it became the catalysts for growth in mainframe applications. These applications largely used a batch model in which transactions are pooled and then applied all at once in an overnight run. Mainframe applications are difficult and expensive to change as well as highly idiosyncratic. These applications did provide business value, and that is why many of them are with us to this day.

Client/Server Era: Distributed Command and Control

In the early 1980s, the personal computer emerged and changed the way applications were built. At first, personal computer applications replicated the monolithic mainframe paradigm in which the application was a single unit performing all of the functionality in a structure chosen by the programmer. But the combination of mainframes, PCs, and networks opened the way for a new paradigm: a two- or three-tier client/server architecture. The client/server era lasted from about 1985 to 1995.

In the three-tier client/server model, a fat client (a PC application) interacts with the user, presenting and collecting information, and then uses a network to interact with a mainframe application that contains application logic to transform data and to store and retrieve data from a database (see Figure 2-2). In the two-tier version, the middle tier is eliminated and most of the application logic resides in the fat client.

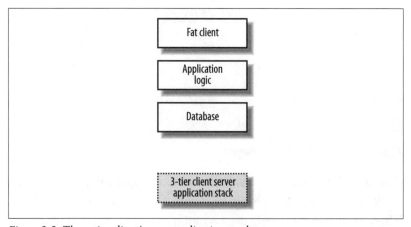

Figure 2-2. Three-tier client/server application stack

This application paradigm, whose adoption was accelerated by the standardization of the UI layer on the Windows operating system and the decreasing cost of desktop computing power, arrived at a time of increasing distribution of management authority throughout globally expanding enterprises. The showpiece application of the client/server era is ERP, which replaced the mainframe applications, performed the financial control applications, and expanded automation to many different areas of the enterprise that surrounded the financial applications. The trend was further driven by the emergence of relational databases, which standardized and productized the database layer, and the increasing proliferation of networks, which provided fat clients access to servers through proprietary protocols and modems.

The problem with the client/server era was the fat client PC application, which drove the total cost of ownership higher because all the PCs on which the fat client was installed had to be individually upgraded. Another problem was the hard-wiring of application logic into the application. Changing the behavior of the application generally meant writing code, then redeploying the application to PCs around the enterprise.

Internet Era: Departmental Command and Control

The Internet era replaced the fat client with the browser, a technology that perfectly illustrated the cycle of technology development. HTML, a standard, paved the way for the browser, a simple product that quickly became a commodity. The browser offered limited application functionality, so all of the application code except the basic user interface logic now moved to the server. Web-based application servers based on common abstractions of the interaction between browser and server appeared. The network was completely standardized, productized, and commoditized based on TCP/IP and related Internet standards (see Figure 2-3). The Internet era started roughly in 1995, with the starting gun sounded by Netscape's groundbreaking IPO.

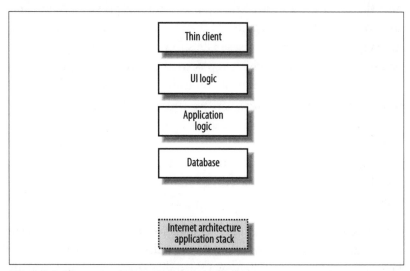

Figure 2-3. Internet architecture application stack

The Internet era continued the decentralization of command and control to the departmental level. Applications sprang up for the supply chain, for sales and marketing, and for procurement. The three-letter acronyms of enterprise applications proliferated—CRM, SCM, and many others—and extended automation to more and more business processes. Business and technology architects saw that important processes such as order-to-cash could be automated across the enterprise applications. XML, a flexible, standardized, and self-describing method for expressing the structure of data, gave way to a new product category, Enterprise Application Integration (EAI), which abstracted the communication between two applications into XML messages. EAI was used to hard-wire automation of processes such as order-to-cash in a pipeline across the enterprise applications (see Figure 1-2 for an illustration). Some of the process would take place in the CRM system, then the information would flow through EAI to the ERP system, and so on. Portal technology, which represented a growth in the abstraction and productization of the browser, helped bring together information from all of the enterprise applications into a single user interface.

In parallel with the rise of enterprise applications, the Internet also brought in the era of collaborative technology like email, discussion lists, chats, web conferences, content management, and knowledge management systems, which automated the exchange of unstructured information. Data warehouses provided a unified view of structured information. The user interface escaped the PC and started showing up in special-purpose devices such as Personal Digital Assistants (PDAs) like Palm Pilots, two-way pagers like the Blackberry, or advanced cellular phones.

The Internet era dramatically improved productivity but also increased the appetite for automation. The reengineering trend of the early 1990s increased the awareness of business processes. Quality improvement programs like the six-sigma framework for continuous improvement amplified the focus on business processes. Corporate executives realized that they were never done with reengineering and that continuous improvement meant continuous tinkering with business processes. But this is where the applications of the Internet era fell short.

The enterprise applications are built in silos, with very little overlapping functionality. To a certain extent, this simplifies the cross-functional process pipelines that are built through them to automate processes like order-to-cash, because each application has a specific job to do. Within the silo of the enterprise application, the behavior can be changed, but the hard-wired integration based on EAI makes changing the cross-functional processes costly and expensive, a matter of coding as opposed to configuring the process. To configure an application means to modify its behavior by adjusting some higher-level description that controls how the application implements a process. Another way to say this is that the ability to change cross-functional processes was not built into enterprise applications.

Corporate executives wanted to bring together the collaborative technologies and the transactional capabilities of the enterprise applications. By transactional, we mean the ability to retrieve, modify, and store a set of records in a database. Portals brought the collaborative and transactional worlds together to a limited extent, but collaborative elements like chats, web conferences, and discussion lists, unstructured information like email and documents, and transactional information were not connected to business processes in any meaningful way. Portals also failed to provide a comprehensive abstraction and unified view of the data and services in the enterprise.

So, as the Internet era begins its next phase, what are we left with along the dimensions we used to start this discussion? Automation has extended to a much larger scope of operational aspects of the enterprise through a proliferation of products. Network standards have emerged as well as standards for many other technologies such as the Lightweight Directory Access Protocol (LDAP) for directories, Rich Site Summary (RSS) for content syndication, and RosettaNet, a group that sets data standards for intercompany communication for many industries. Application servers have become productized and standardized to some degree and are well on their way to becoming a commodity. The abstraction of the application stack is now on the verge of adding more layers. The application and UI logic that sit between the browser and the database are about to add a layer of process or business logic that will allow for increased flexibility at the process level. Vendors of all stripes are now trying to lay claim to this new layer and lead the next wave of IT evolution, which we argue is the PCA/ESA era.

The PCA/ESA Era: Employee and Team Empowerment

The technology of the PCA/ESA era will break through the limitations of the Internet era and extend automation across all departments, divisions, and functional boundaries to the employee and team level. To accomplish this, processes cannot be bound within the silos of existing enterprise applications or limited to rigid pipelines that flow through them. Designing, automating, and modifying processes cannot be expensive. A business analyst or application architect must be able to look at a unified view of the enterprise, decide on the processes and functionality needed to automate a business task, and then rapidly build or buy an implementation. Productization, standardization, and abstraction will all take place on this playing field, and commoditization will continue to take in new layers of technology.

PCAs are the thin end of the wedge for this new era of technology because they offer the needed functionality at the lowest cost by building on existing systems without disrupting them. PCAs are made possible by an Enterprise Services Architecture (ESA), the blueprint for how IT infrastructure will be organized in the future. ESA platforms will provide the technology that will allow PCAs and custom composite applications to be constructed on top of existing systems (see Figure 2-4).

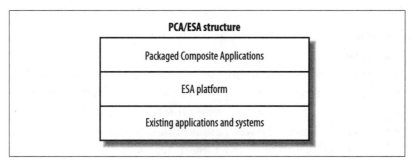

Figure 2-4. PCA/ESA structure

The next sections discuss how the PCA/ESA era has emerged so far and will continue to play out.

Ubiquitous Services

The first change that marks the arrival of the PCA/ESA era is the universal adoption of the concept of services. A service is an abstraction of a unit of functionality that can be invoked in a variety of ways and is loosely coupled to the programs that invoke it. By abstraction here we mean a description that hides all of the complicated plumbing underneath. Loose-coupling means that the service doesn't need to know much about who is invoking it, and the program or user invoking the service doesn't need to keep track of what's going on inside the service.

A great example of a service is a request for a web page, which is identified by a URL. A web browser requests a URL, and the web server responds by sending the page or an error message if it can't find the page. The browser and the server speak HTTP, the protocol or language they use to communicate. Usually the page is in HTML format, but it doesn't have to be. The dramatic growth of the Internet has been fueled by the simplicity of this architecture. The browser and the web server are loosely coupled because they don't have to know much about each other to interact and, once their interaction is over, they don't have to keep track of each other. Both programs speak HTTP, and that is enough to carry on their conversation.

The notion of a service, even a loosely coupled service, is not new, but the success of HTTP was unprecedented. This success led to the idea of generalizing the notion of a service, which is exactly what web services protocols do. The simplest way to understand web services is as a generalized HTTP-style protocol that can be used for all sorts of conversations between applications. (Some web services use a fancier protocol called Simple Object Access Protocol, or SOAP, to carry on their conversations.) If XML extended the idea of HTML into a general language that could describe the structure of data, then web services extended the idea of HTTP into a general way of describing and invoking services that preserves simplicity and loose-coupling. Like HTTP, web services are not bound to a specific operating system or computer language. Like HTTP, they are not difficult to implement. Like HTTP, implementations of the web services protocols will be quickly commoditized.

The focus on web services and the simplicity and power of the technology allow us to assume ubiquitous adoption. The typical standards battles are being fought, but it is only a matter of time before all applications and systems speak web services as a primary mode of interaction. Both corporate IT departments and software vendors of all types are exposing their application functionality through web services.

Integration and Modeling

So, in the PCA/ESA era, everything will be a service. Hooray, but why should I care? To answer this question we must look at why integration of the process pipelines mentioned earlier was so expensive and difficult to change and what can be done about it. In the Internet-era applications, automating a cross-functional process meant creating a pipeline of information and functionality through all systems involved in that process: ERP, CRM, SCM, and so on. But what did it actually mean to automate that process and integrate the applications?

Generally, pipeline integration was initiated in response to competitive pressure. Either a competitor was gaining an advantage through increased efficiency or a company realized it could gain an advantage over the competition. For example, if the time to process and fulfill an order was cut from ten days to five days, all sorts of potential benefits might accrue. With the desire for more efficiency in mind, a technology architect and business analyst would set out to examine all the steps needed to fulfill an order, and generally they would find inefficiencies like manual reentry of data, delays in the process while information was assembled, management approvals that were needed because the line employees didn't have enough information to do their jobs, and so on. The technology architect and business analyst would then create a chart of how the process should flow in a graphical program like Visio and then map out the integrations that would improve efficiency.

Integrations usually involve moving data from one enterprise application to another. This is where things get difficult and expensive. Integrations are difficult not because it is hard to get information out of an application or to put it back in. That is not a problem for most modern software. The tricky part of integrations is having the information mean the same thing after it is transferred—that is, getting the semantics right. Different applications use different coding schemes, different ways to represent quantities like currencies or interest rates, different ways of recording addresses or accounting categories. To reconcile the different semantics, a model must be created that allows the information from both sides to be mapped into one structure that does not leave any important information out. This is sort of like converting Eastern Standard Time and Pacific Standard Time into Greenwich Mean Time. Once the model is constructed, integration involves converting information from both applications back and forth to the model.

Changes to these sorts of integrations are expensive because the conversions to and from the model are generally tightly coupled and hard-coded. Further, the models are generally poorly documented. Application Programming Interfaces (APIs) provided by the vendors of the enterprise applications are the vehicle used to implement most pipeline integrations. APIs get the job done brilliantly; they move the XML messages back and forth between applications, but they do so in a tightly coupled manner. The program orchestrating the integration between the two applications must know a lot about the internal state of the applications on either side of the integration. This makes the integration complicated and expensive to maintain and upgrade. In order to fix anything or improve it, the developer must know a great deal about how each application works and about the model that brings together the information from each application. Hard-coding means that the integration is implemented in a programming language, which makes it difficult to change because expert programmers are required.

Complexity in these sorts of integrations multiplies rapidly as the number of point-to-point connections grows. Maintenance expense also increases with the number of point-to-point connections. Every once in a while this sort of complexity is simplified by the creation of a standard model and interface to a key application that all other applications must accommodate.

This approach to integration is expensive and hard to change, but it actually works. The pipeline processes do indeed improve efficiency in general and provide a competitive advantage in the best cases, which makes a powerful case for further automation. But the inflexibility of pipeline integrations causes problems down the road and the hard-coding and tight-coupling make such integrations expensive to maintain.

The ESA Platform

With the rise of services and the expense of integration in mind, we can easily get a clear picture of what an ESA platform does to enable PCAs to be created. The ESA platform constructs a universal model of enterprise data, services on top of that data, and processes that describe sequences of steps and the services and data involved in each step. (We frequently refer to data and objects interchangeably in this discussion to emphasize that data structures are tightly bound to certain basic services. Objects are data plus services that are closely related to the data.) The ESA platform constructs this universal model out of the objects, services, and processes provided by the existing applications and systems in an enterprise. This is an idea with profound implications.

One of the most valuable aspects of an ESA platform is productization of integration. To create a master model of the objects, services, and processes of an enterprise is no small undertaking. Mapping each application to that master model is also a daunting task. But once done, it creates a platform of awesome power and flexibility. One can think of such a platform as all of the applications in an enterprise with all of the possible point-to-point connections made.

The master object model in an ESA platform brings together all of the information about the various entities that enterprise applications manipulate into one unified form. A master model would include an object for customer, another for employee, another for business partner, and another for supplier. Relationships between the objects to express hierarchies like organizational charts are captured in the master model. The objects in such a model are created from data and services provided by the existing applications. Implementing the mapping between the enterprise applications and the master model is one of the primary tasks of the ESA platform.

On top of the objects of the master model in an ESA platform is a unified set of services that manipulate these objects. Some services are part of the object itself. A customer object might have services that return the last name, full name, or address. The higher-level services that the ESA platform unifies from the existing applications perform functions such as sorting customers, searching for a customer by social security number or tax identification number, or finding all the accounts for a customer across divisions. This abstraction of services reduces the amount of tight-coupling required to use objects and services from the existing systems in an enterprise, and it creates a single method for accessing certain objects. In other words, it creates a standardized view of the enterprise. For this reason, it is quite possible that the emergence of ESA platforms will set off a new wave of standard setting, an issue we discuss in Chapter 4. Another benefit of this unification is that it brings the transactional and collaborative worlds together so a developer can use functionality from both to develop applications against a common set of data.

In the discussion earlier about the way programs evolved we noted how the logic or programming that performed the automation and did the needed work has started to form into layers. In the mainframe era, it was all one layer. In the client/server era, two layers were formed, one for the user interface and processing on the client, and one for the application logic on the server. In the Internet era, the amount of logic on the client shrank to simply handling user interface logic, with the application logic on the server doing everything else.

Tight-coupling occurs in the application logic. This layer has to know everything about the underlying systems and the user interface and sort it all out so that automation takes place. Application logic performs many different sorts of work, including:

- Communicating with databases and networks
- Manipulating and transforming data from one layer to another

- Organizing data for presentation to the user interface
- Sending data back and forth to the user interface

Somewhere in this application logic, the business process logic is also handled. By business process logic, we mean the knowledge of what stage we are at in a process, how many steps within that stage have been accomplished, and what the next stage will be, based on how the work proceeds.

Most enterprise applications have processes hard-wired into them. Some have the ability to configure these processes, perhaps even with a graphical interface. Standards for describing processes also exist, but there is no way within current enterprise applications to describe processes that cross the boundaries of enterprise applications.

The ESA platform reaches into the application logic layer and pulls out all of the process logic from each application. The ESA platform creates a standard description of processes across all enterprise applications and a standard way to manipulate and configure them. The functionality and processes in the existing applications are used to implement some of the standard processes and the objects and services can be invoked or linked to various stages of the processes. A process for creating a customer, for example, might involve collecting basic data. The second step could be looking through the existing systems for that customer and identifying potential duplicates. The next step could be creating the objects needed to process orders from that customer in the underlying systems. The final step could be handling the order that started the whole process. A thorough process description will have exception conditions defined at each step and further processes defined for resolving those exceptions, automatically or with manual intervention. Figure 2-5 shows the separation of the process logic in the PCA/ESA application stack.

Ideally, an ESA platform builds on the graphical description of a process created for the tightly coupled pipeline integration. From that starting point, the process can be tweaked and configured as needed. The abstraction of the business process layer from whatever mess

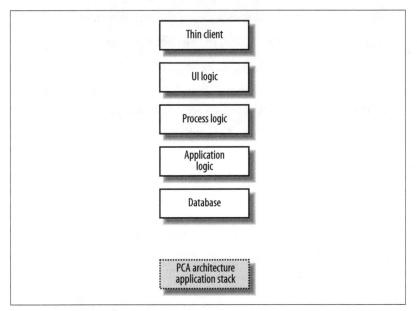

Figure 2-5. PCA/ESA application stack

remains in the application logic enables business analysts to tinker and refine business processes that are not bound within the silos of the enterprise applications. Configurable business processes can then orchestrate unified objects and services, all of which constitute a packaged integration of enterprise applications that reduces the cost of changing the infrastructure and allows for rapid optimization and automation of the enterprise.

This approach has tremendous implications for how software will be developed in the future, which we will discuss in Chapter 4. It also sets the stage for standardization of various enterprise objects and services as well as rapid development of PCAs.

The Power of PCAs

The fundamental goal of PCAs is to extend application functionality throughout the enterprise to the greatest number of stakeholders. But to do this we need to change the way we build software, and that, and economic forces, are the primary drivers behind the creation of the ESA platform and the PCA mode of application delivery.

Wouldn't it be great if the ESA platform could just be willed into existence? In many ways, the ESA platform represents the unification of a variety of product areas. Products in the Enterprise Information Integration arena are aimed at creating a unified model of corporate data. EAI products, which started as systems to pass around XML messages, are being retooled to add support for web services. Business Process Management (BPM) products allow the definition and modification of processes and the ability to connect them to existing data. BPM layers are being added to application servers, EAI systems, and other products. The fundamental notions of a variety of other modeling and integration technologies are included in an ESA platform.

But we cannot, through force of will, incarnate a fully functional ESA platform that fulfills the entire vision of ESA, and, to be honest, we know from experience that if we could do so, it would probably be wrong. ESA platforms will arrive in stages. The first generation of ESA platforms will arrive in 2003 and will implement a large portion of the ESA vision. We examine what sort of functionality will appear at each stage of the ESA platform's evolution in Chapter 4.

The ESA platform may be the first leap forward in IT infrastructure that is being pursued from a perspective of humility, despite its revolutionary potential. There are few senior architects who still believe in the waterfall process of development, in which a project is defined in detail in its early stages based on requirements gathered from the field, then implemented and launched. Most software architects now believe in an iterative approach in which software is created and launched and then gradually revised toward perfection. The fundamental axiom of the iterative approach is that requirements are not fully understood until a working system is in the hands of users. ESA platforms are being constructed in stages, partially because of the influence of iterative development methods, but primarily because the vision is too large to implement all at once.

The question, then, is how can a corporation benefit from a partially implemented ESA platform? The answer is Packaged Composite Applications.

PCAs are software products that will be enabled by the first generation of ESA platforms. The ESA platform delivered to support each PCA will have enough of the master model and the unified objects, services, and process layer to allow the PCA to do its job on top of a subset of existing applications. As time passes and a company buys more and more PCAs, the ESA platform will grow. After a year or two, the ESA platform will be fully implemented with a unified model encompassing all enterprise applications.

As the ESA platform matures toward the point of being a development platform open to all developers like an application server, PCAs developed by vendors will solve the problems outlined so far. PCAs are products, so they have an attractive cost of ownership. They are built to accommodate flexible and configurable business processes. They combine collaborative and transactional functionality in a way that enables more complete support for strategic processes. They are focused applications with a modest scope that solve urgent problems without disrupting existing systems. They leverage existing technology and bring the functionality and information from existing applications to a wider audience. And PCAs can automate processes across the traditional silos of enterprise applications.

The result is that enterprise architecture becomes more flexible, not in a big bang but in a gradual way, starting with PCAs that address the most important points of pain, then through more PCAs, which complement the first generation, and then finally to the point where the fully realized ESA platform becomes the foundation for all development.

Another type of humility is at work here, one driven by economics. The PCA/ESA paradigm would be doomed to failure if it were expensive or required a wholesale replacement of applications and infrastructure instead of leveraging the investment in existing technology. Vendors are leading this revolution with PCAs because they are smaller in scope, affordable, and deliver a quick return on investment while solving precise points of pain. This is a sale that can be made in a challenging environment for IT spending. Admittedly, an ESA platform is a more difficult sale, requiring substantial evidence that it will save money or increase revenue. Vendors also realize that anything they sell must leverage the investment in existing technology, which was the focus of a huge investment to remedy the Y2K bug.

So far we have talked about the shape of the PCA/ESA infrastructure and how it will evolve. In the next section, we look at the connection between technology and economic reality. We ask some prominent individuals in the field to examine the question of whether the sort of functionality that PCAs can deliver is exactly what the enterprise needs from its technology infrastructure and what companies should do with their technology in order to succeed and thrive.

What the Enterprise Needs

The promises of technologists to senior management have been broken all too frequently. Technologists imagined systems that would transform business and the world, ushering in a new age empowered by technology. It rarely worked out that way.

What happened instead was the steady march of automation that brought about dramatic improvements in productivity and paved the way for new strategies that did indeed transform industries. Perhaps now that technologists and senior management have struggled together for a generation, they are on the same page and can move forward with realistic expectations.

To carve out the vision for the next step for enterprise IT, we have sought out several leaders in the industry to provide perspective from the vantage point of the customer, the CEO, the investor, and the analyst. We will use these viewpoints to provide the foundation of our arguments for PCAs.

The Customer Perspective

Alan Cooper, the father of Visual Basic and a pioneer in the field of interaction design, frames the problem facing the enterprise by focusing on the computer systems that are increasingly intermediating the contact with the customer.

The problem for enterprise IT, from Cooper's point of view, is that the touchpoints with the customers are frequently making them angry. We have all been trapped in circuitous interactive voice messaging systems and unable to talk to a human being. We've all had the experience of being asked to provide the same information repeatedly, commonly through a phone keypad and then again when we finally speak with a person. Computer-mediated interactions with customers are marked more by failure than by success.

In *The Inmates Are Running the Asylum*, Cooper catalogs design failures in common interfaces and explains why they occur. He points out a large gap in the skills of the technologists, who are charged by default with designing these interfaces and interactions. The technologists, whom Cooper calls Homo Logicus to emphasize how differently they think, tend to see user interfaces in terms of the way the underlying program is implemented. That is why so many tree structures, hierarchies, and other programming concepts show up in user interfaces. Cooper says the correct way to design interfaces is to focus on how the interface will meet the goals of the users. He sets forth a method for doing this based on personas, a complete understanding of the different people who will use a system.

Good design, Cooper points out, is one of the most compelling competitive advantages. He credits the survival of Apple Computer to the fanatical devotion of its customer base, a devotion fueled primarily by excellence in design. In his book, Cooper compares user perceptions of programs with simple interfaces and fewer features with programs having less well-designed interfaces but more functionality. Users consistently rated the programs with simple interfaces as more powerful.

Cooper concludes that investing in good design speeds development, makes users happy, and provides a sustainable competitive edge.

Besides hiring competent designers, one important way that applications can be made more powerful is by providing those designers with exactly the right application services to help the end users meet their goals. It is futile to design an interface to display all accounts summarized on one page if the desired information cannot be pulled together at a reasonable cost. By leveraging the power of existing applications, the PCA/ESA paradigm offers the interaction designer more potential functionality to choose from. The result in the hands of a competent designer should be a better user interface at a lower cost.

The CEO Perspective

James D. Robinson III was CEO and Chairman of American Express for 16 years. He sits on the boards of Novell, Coca Cola, and Bristol Myers, and is a founding partner in RRE Ventures, a venture capital firm that invests in information technology companies.

Robinson's career spans the entire evolution of the IT infrastructure. The mainframe era, in Robinson's view, resulted in significant automation that provided more information for senior management. But the budget for Management Information Systems, as IT was called at the time, kept climbing ever higher and the systems were generally slow to arrive and delivered far less than they promised. The client/server era essentially repeated this cycle of disappointment, replacing the mainframes with a new more flexible architecture and replacing the central mainframe applications with ERP systems, whose implementation was a massive and expensive undertaking for most companies.

In the Internet era, the fear of disintermediation and of new competitors like Amazon on the one hand and the push to fix the Y2K problem on the other drove a massive investment in IT. Automation was extended to applications like CRM and SCM, but, from Robinson's perspective, the benefits were muted because the information was trapped in the silos of each application. The expansion of global trade during this period and the productivity derived from IT investments drove high growth rates with low inflation until the bubble burst. Then, with the Y2K problem resolved, the CIO was put in the penalty box for having spent a tremendous amount of money without sufficient return. Since Y2K, IT spending has been sharply curtailed. As a result, the larger vendors had the upper hand and were able to regroup as the smaller companies struggled to survive.

The problems Robinson sees embedded in the current infrastructure are rigid systems that keep information from being combined in ways that would help better run the business. "At many companies you will have customer interface points in several places that will never know what's going on in other parts of the company," Robinson points out. "It is amazing how much investment has taken place and how many systems have been implemented, yet companies are still not able to get the information tied to the financials, tied to the operations, in ways that provide a comprehensive view to see where goals are being reached, where they being missed, how to divert resources from one area to another."

According to Robinson, one of the key challenges he sees facing CEOs and CIOs is preventing the organizational, functional, and operational divisions from getting in the way of running an efficient business. Robinson strongly suggests that companies forget about resolving the tensions between centralized and decentralized structures and forms of control and instead pay more attention to creating systems that can assemble the information needed to run the business.

His advice to CEOs is to have the CIO as a direct report and to be wary of technologists who want to build rather than buy solutions.

His advice to CIOs is to stay close to the needs of the business and find ways to empower line-of-business executives through IT.

Despite his eyewitness view of IT's failure to meet expectations, Robinson remains an enthusiastic advocate of IT as a strategic asset.

He sees technology improving over time; for example, one of his previous companies tried to unify more than 30 corporate master databases, a massive project that ultimately failed. However, the project would succeed now because the tools for data profiling and unification have evolved.

He believes any future architectural retooling will have to be evolutionary and build on the current generation of IT systems. "Nobody's going to take out the existing infrastructure," Robinson says. "There's too much sunk cost in there."

His view is that better management information will increase the power of central management control and reduce politics. The key to getting things right in his view is more detail, getting down to the individual employees and customer view. Every year at most companies, a budget is devolved into departmental goals and then into goals for each individual in the company. If the systems of a company allow senior management to see at the individual level what is working and what projects are late, and then roll this up into what parts of the plan are at risk, intelligent decisions can be made in time to save millions of dollars.

"The companies who are going to be the survivors are those who know how to utilize IT intelligently to drive better customer service, better productivity, and more efficient delivery of the entire value chain and that can be a big company or a small company," Robinson says. "It is not a race between the big and small but between the guys that get it and the guys that don't."

The Investor Perspective

Dr. Nancy Martin, PhD, is a partner at Warburg Pincus, a private equity firm that makes significant investments in companies in all stages of development. In her career she has observed the evolution of the IT infrastructure from the mainframe to the Internet era from a variety of vantage points. She has founded her own company, been a partner at a Big 5 consulting firm, run R&D for a Fortune 100 corporation, and served as a visiting professor at Stanford. In her role at

Warburg Pincus, she evaluates investments, participates with portfolio companies as an adviser and board member, and keeps a close eye on how the IT industry is evolving.

Martin's analysis of the IT infrastructure starts with the observation of how companies pursued client/server technology for the promise of flexibility, faster response times, better usability, and more functionality, but in doing so actually took a step backward in several respects. In the mainframe era, business units were frustrated with the pace with which IT met their business requirements, and the first wave of client/server adoption was driven by executives putting in client/server systems on their own, side-stepping the bottleneck of the central IT organization. The new applications, which worked fine at the departmental level, lacked the security, performance, reliability, and availability of the mainframe applications. There was also a proliferation of technologies without agreed-upon standards. When IT reasserted itself and the client/server applications were brought under central control and management, deficiencies in the technology that were manageable at the departmental level were more acute at enterprise scale.

Martin's view is that although automation was extended to new functions in the client/server era, the applications did not live up to their promise and their narrow focus resulted in the creation of distinct silos of data of the sort that never existed in the mainframe era (of course, neither did most of this type of data). In the Internet era, usability suffered further, as the browser replaced the fat client in pursuit of increased availability and the desire to decrease the total cost of ownership by lowering the amount of maintenance at the desktop. Browsers are less costly to install and maintain than fat clients. The current phase of development involves returning to and exceeding mainframe levels of quality. "What we're doing now is combining things we lost from the mainframes with things we never gained in client/server," Martin says.

Martin is wary of offering general advice to CIOs, but her advice to portfolio companies is guided by several enduring principles. When she visits a company, she hopes to find a technology strategy that is tightly aligned with the business strategy and a set of tactical plans for executing that strategy. Investments in technology should be pursued only for business benefits and business needs, not in the pursuit of an amorphous need for more functionality or the current fad in technology. Companies must know when they should be conservative and when they should be aggressive. "Different businesses need to be at different places in the technology curve," Martin says. "Some businesses need to be at the bleeding edge, at the forefront of technology, and other companies should only adopt a technology when it is proven. There is not one piece of advice you can give about new technology because companies' perspectives are different. Know where you are on the curve. Otherwise, you can waste a lot of time and money or you can miss opportunities."

Martin sees several potential pitfalls for CIOs given the current trends in the industry. She sees a clear need for companies to undo the barriers to appropriate access to data, barriers created by applications that locked data into distinct silos. She sees a strong case for increasing the flexibility of the IT infrastructure and for supporting processes that span departmental divisions. The move to a service-based architecture from the bottom up has happened in some very progressive companies but is still unknown in many others. However, progress forward, she believes, must take into account the current plethora of systems and solutions in the infrastructure.

"We're still using applications that were built 30 years ago for which we don't have the source code," Martin says. "So it's clearly true that there is functionality that has been defined and embedded in corporations that we can't extract and that's going to continue. We're not going to throw away our existing applications for the most part."

Martin is wary of quick fixes or easy answers. Throwing web services at an infrastructure won't automatically increase interoperability without solving other key problems such as the creation of a common data model. Existing applications will be hard to turn into services without knowing the side effects entailed in using existing functionality. Creating flexibility for business processes would be wonderful, but she cautions CIOs to be skeptical of vendor claims that flexible business processes can be achieved quickly without significant preparation. Service-oriented architecture might be a significant advance, but only if proper attention is paid to where services will execute, how they will be defined, and what granularity will be required.

She suggests that CIOs look at the history of why projects and efforts succeeded or failed in the past for guidance. "The biggest thing you should be careful of is looking beyond the buzzword and seeing the reality," Martin says.

The Analyst Perspective

In his analysis of the evolution of IT infrastructure, Dr. Ravi Kalakota, PhD, sees a cycle of centralization and decentralization caused by the interplay between organizational structure and the capabilities of technology. Kalakota, who has been a professor and an entrepreneur, is CEO of E-Business Strategies, a technology research and consulting firm. He has written several books on e-commerce and e-business.

Kalakota's observation is that business technology reflects organizational orientation and business drivers. In the 1960s and 1970s, the largely centralized organizational form was the rule and the mainframe met that need perfectly. Three things happened to change the mainframe emphasis. First, users became frustrated with central control and the lack of responsiveness to their needs. Second, client/server architecture allowed users to do much more for themselves without consulting IT. Third, a shift in management philosophy occurred: IT was viewed not simply as a cost center but as a revenue source. This paradigm shift largely parallels the rise of the Web. In Kalakota's view, the popularity of both client/server and the Web sprang from their support for decentralized organizations rather than from their technical superiority.

"We are going through a sinusoidal pattern from centralization to decentralization and back," says Kalakota. "Companies centralize to get economies of scale, [which] disenfranchises the user. Then technology shifts and gives the user power that results in decentralized activity, but the backend cannot adapt. To control the chaos and bring about better integration, managers centralize again."

In the late 1990s, the twin imperatives of cost control and growth became the driving force behind IT investment, and that was reflected in the enterprise applications of that era. Now, in the lean years of the new millennium, competition has grown because the Internet has provided the consumer with more information than ever before. Simultaneously, cost pressure is high and revenue is hard to come by. Companies are focused on satisfying the customer in every way possible: providing more service, providing faster service, cutting prices, reducing expenses, and customizing the product. Doing whatever is required to keep the customer is the order of the day.

Using an analogy from microbiology, Kalakota compares business problems to a virus and technology to antibiotics. The overuse of antibiotics often results in viruses that adapt and evolve. Similarly, when a problem first appears, the technology is prepared to meet the challenge, and generally after some false starts, the business problems are solved. But then just like a virus, the business problems morph and require new solutions that may go beyond the scope of the existing technology. New technology is required to solve these problems.

While the problems being solved shift every three or four years, Kalakota believes that two important paradigm shifts are in process that should receive full attention from CIOs.

The first paradigm shift is in the nature of automation. Up until recently, the CIO looked inward for her to-do list. The focus was on task automation, making the operations of the enterprise as efficient as possible. Now the problem has been turned inside out. The enterprise has become a service or a collection of services. The to-do list comes from the customers. Customers don't care what's going on inside the service; they want fast response, significant value, and an affordable price.

"In order for companies to survive they have to digitize their business and be adaptable to customer needs," says Kalakota. "The pressure to deliver the services (via brick, web, tele, and mobile) that the customer desires and cares about is increasing."

The second paradigm shift is in how IT is gradually yielding to the same pressures that caused massive changes in manufacturing over the past 30 years. At one point, a manufactured product like an automobile was constructed from thousands of parts. Now, in response to pressures to continuously reduce costs and provide more customization, automobiles are assembled from 20 or 30 component subsystems. The thousands of parts still exist, but they are encapsulated in subsystems that fit together in standardized ways.

Most CIOs still see their businesses as a large connected set of relatively small moving parts. The complexity inherent in this structure makes organizations difficult to change because there is no encapsulation. The prospect of upgrading all but the most trivial system is quite difficult in a fragmented infrastructure that is not componentized because the number of connections between systems is large and change may affect many different systems.

The high cost of change—or aligning business and technology—works against the business in many ways. The company cannot take advantage of improvements to commercial software in their infrastructure because upgrades are traumatic. The cost of changing the infrastructure to respond to new problems or customer demands is higher than it should be. Perhaps the biggest problem is the difficulty of understanding how and where to improve a system constructed from a multitude of parts. The effect of a change is hard to predict and the cost of understanding the system and keeping that understanding current is high.

According to Kalakota, "The job of the CIO is changing again. In the early 1990s, it moved from data processing to business process reengineering (BPR). In the late 1990s, it moved from BPR to e-business. In the early 2000s, it is moving from e-business to e-services. The next phase of CIO evolution is pretty clear: multi-channel ser-

vices—the design, implementation and delivery of integrated
vices." Without an adaptable, componentized infrastruct
deploying multi-channel services is going to be a costly and time-
consuming exercise.

The CIO To-Do List

If the viewpoints we have just discussed represent what the CIO's
clients want, what then must be done to meet those demands? In
this section we will examine the contours of a practical answer and
the specific challenges that will face CIOs at most companies.

Imagine the CIO, surrounded by the people who make up her con-
stituency. The customer wants better service. The investor wants
more return for each dollar spent on IT. The CEO wants integrated
systems and better information faster. The CFO wants to push down
costs. The line-of-business executive wants more automation and
flexibility. While past failures have created significant skepticism,
every member of this group is still looking to the CIO for answers.

The evolution of IT infrastructure and the current economic reality
give the answers certain boundaries. About 70 percent of IT budgets
supports existing technology, and that does not include the increasing proportion of the remaining 30 percent that is focused on integration. There is not a lot of money sloshing around for new IT
projects. The scarcity of resources may become the norm for IT. The
question facing IT and business managers now that cost-cutting has
run its course is how to innovate during a drought.

Make Incremental Progress

One characteristic of the next stage of evolution of IT will be incrementalism. Smaller projects with smaller budgets that are faster to
implement will be the order of the day. Large initiatives may occa-
sionally be funded, but only after the risk has been dramatically
reduced by experience gained from smaller projects. Changing every-

thing through a big bang replacement of current infrastructure is out of the question.

Drive Down Infrastructure Costs

The economy will ebb and flow but the demand to cut costs is likely to remain constant. IT spending at most companies has become a huge portion of the budget and this line item will continue be scrutinized closely by the board of directors. Increasing commoditization of technology and competition from open source technology will keep pressure on technology vendors. Senior management will demand that CIOs reap the benefits of these trends for their companies. Vendors will respond with more and more commodity products to replace more expensive infrastructure and increase productivity.

Reduce the Cost of Change

All the experts agree that we must lower the cost of change. IT architects are going to have to build flexibility into their systems at all levels to allow a much more rapid cycle of innovation. The pace at which a company can optimize and innovate will be a function of the cost and speed of changing current infrastructure. Companies who are hamstrung by an inflexible complex of systems will be at a competitive disadvantage.

Leverage Existing Infrastructure

Incrementalism and cost pressure mean that current systems must be the foundation for progress. Successful CIOs will find a way to decrease costs and increase flexibility with each incremental step forward.

Extend Automation

Automation speeds the velocity of the enterprise. Instead of launching five products a year, a fully automated company could potentially launch ten. Given how the enterprise applications have focused on automation within departmental and functional silos, the next wave of automation will focus on processes that cross the traditional boundaries of those applications.

Opportunities also lie in automating strategic processes. Strategic decisions can occur faster because the information on which to base those decisions can be assembled and distributed faster.

Provide a Unified, Real-Time View of the Enterprise

A unified view of information is needed to support decisions and monitor progress at all levels of an enterprise. The mere presence of this information can be a transformational force that reduces politics and improves the performance of teams.

Possible Alternatives to PCAs

The argument of this book is that the list of tasks in the previous section is best approached by a continued focus on finding PCAs that meet these needs and gradually adopting an ESA platform to unify objects, services, and processes.

In Chapter 3, we examine the arguments for and against this position. To complete our clearing of the decks in preparation for this case we will examine the other approaches that could be taken by a CIO to achieve the goals outlined above.

The possible alternatives to PCAs range from doing nothing to replacing everything. In this section we examine the strengths and weaknesses of each option.

Maintaining Business as Usual

One plausible way of reacting to the competitive challenges would be to avoid significant changes and incrementally improve performance based on existing technology, applications, and methods. While this approach may seem attractive in the face of so much over-promising and under-delivering from technology in the past five years, it also ignores the inexorable march of automation that is taking place in every industry. The benefit of this approach is that it is not traumatic and eliminates new investment. The costs of this approach are measured in the decay of competitive position that may occur if a company falls behind in technology, and in any lost savings or efficiencies based on delaying further automation. In the rare industry where technology does not provide an advantage, this could be a reasonable course.

Starting from Scratch

The other extreme is hitting the reset button and rebuilding the entire infrastructure to maximize flexibility and reduce costs. But few people believe that such "big bang" approaches have a chance at success, and, further, there is often no money to fund such a massive undertaking. It is not reasonable in the current environment to expect companies to abandon the investment in current infrastructure. It is not clear what sort of companies would find this approach attractive.

Integrating Existing Systems

A wealth of products dedicated to integrating existing systems has arrived in the past five years. Enterprise Application Integration vendors offer to connect applications with messaging infrastructure. Business Process Management vendors provide a way to knit together systems exposed with EAI. Web services vendors offer to expose applications as web services and orchestrate their behavior. Enterprise Information Integration and data warehouse vendors offer to bring all corporate data together into a unified repository. XML vendors offer to encapsulate everything as XML and bring it all together. Why can't all this plumbing help improve strategic efficiency and implement cross-functional automation?

The short answer is that it can, but the costs quickly run high. The first problem is that most of these approaches are focused on knitting together existing systems, not on adding new functionality. They fall short of the PCA vision because they don't provide a place to create new services and objects to automate new tasks. The second problem is that these products must be licensed and integrated. Lots more money will be needed. The final problem with integration is that it creates a new layer of custom development to maintain.

It is fair to ask, however, how the PCA vision differs from integration. First, in one product, the ESA platform vendor brings together all of the functionality that would otherwise be integrated through separate products, such as EAI and data warehouses. Second, specific business solutions are offered and supported as products. In essence, integration amounts to do-it-yourself PCA and ESA.

Extending Existing Systems as an ESA Platform

Another plausible alternative for some companies will be to extend an existing large platform, such as an ERP system, to provide more strategic efficiency and cross-functional optimization. This approach avoids the license fees of integration but has two failings. First, it creates more custom development that must be maintained, and second, it creates a do-it-yourself approach to integration with enterprise applications that will be productized by ESA platform vendors. However, if one large application solves 80 percent or more of a company's needs, this may be the way to go.

End-User Collaboration Tools

From email to shared folders to desktop productivity applications like word processors, spreadsheets, and presentation programs, end-user collaboration tools have fundamentally changed the way we work. Why can't these tools provide the solution to strategic efficiency and cross-functional optimization? We already have them and know how to use them.

This is a powerful argument. For some parts of the solution, these tools are essential, and, as we argue in Chapter 3, PCAs will fail— any modern applications will likely fail—without appropriate integration with end-user tools.

But look at the reach of the end-user tools. They reach to your desktop and to the desktops of other people, but they do not reach into the enterprise applications that contain the data most important to increased automation. A unified view of corporate information is vital to achieving new levels of automation. That cannot be provided with desktop tools.

Even when it is possible to extract information from enterprise applications, the transfer is generally one-way, providing a read-only view. End-user tools do not allow people to reach back into an enterprise application and take action. They don't allow integration of the functionality of several enterprise applications into a new function. Finally, they don't allow extension of the functionality of existing applications.

Now that we have looked at history, examined how the PCA/ESA paradigm represents the next step, heard from the experts, and looked at the alternatives, we are ready to dive into the cases for and against PCAs, the focus of the next chapter.

3

The Case for PCAs

Short of cash, with sharks circling in the marketplace and costs cut to bare bones, companies are operating at high levels of efficiency with their current infrastructures and processes. If they aren't, their competitors are beating them about the head.

Our examination of the evolution of IT infrastructure over the past 40 years showed us how enterprise systems have gradually evolved from supporting financial and control applications, to ERP, and then to best-of-breed enterprise applications like CRM and SCM. The next frontier is applications that allow for flexible and configurable cross-functional automation of processes.

So how do we create such applications? In Chapter 2 we examined several unsatisfying options. Doing nothing is one end of the spectrum. Starting over and rebuilding a new infrastructure is the other. In between are choices like creating composite applications with custom development by integrating current systems, using one system as a *de facto* ESA platform, or making do with end-user collaboration tools like email, spreadsheets, and shared directories. Some of these options get part of the way there, but each has major flaws.

But why then are PCAs the answer? In this chapter, we examine the arguments for and against PCAs. Presenting all of the arguments is important, not because we are trying to win a high school debate and persuade a panel of judges that PCAs are right for everyone, but because the most important question for readers of this book is, "Are

PCAs the right choice for me?" Surveying the arguments, pro and con, is the best way to properly answer that question.

Arguments in Favor of PCAs

Why should anyone look to PCAs as an important part of their future? The following sections aggregate the arguments for PCAs under five propositions:

1. PCAs leverage existing investments

2. PCAs increase strategic efficiency

3. PCAs enable cross-functional optimization

4. PCAs reduce costs

5. PCAs manage change

First, let's look in detail at how PCAs leverage existing investments.

PCAs Leverage Existing Investments

Unlike the other approaches described in Chapter 2, PCAs are not a development effort. They are products that are installed and configured to bring a combination of new and existing functionality to a focused business process. In this way, the unique structure of PCAs amplifies the value of existing systems. Because they are based on the ESA platform, which allows PCAs to snap onto existing systems through a productized integration, PCAs don't disrupt existing applications. Let's delve deeper into these two areas.

Existing systems are not disrupted

The ESA platform allows PCAs to communicate with existing enterprise applications without disruption. We will explain in detail how the ESA platform achieves this in Chapter 4. Briefly, however, here's the high-level story. The ESA platform combines two sorts of systems to create an environment for developing PCAs. The ESA platform has adapters that move information back and forth from enterprise applications like CRM and ERP. The ESA platform also uses functionality from platform component systems like content management, business process management, portal systems, and the like to manipulate the information from the enterprise applications. One of the key methods to communicate with applications is through web services.

The simplest sort of service for an enterprise application allows PCAs to read data. In the middle ground are services that allow reading of data and some sort of transformation on data from the existing application. Full support consists of allowing data to be written back to the existing application. As much as possible, the ESA platform handles translating messages between the PCA and the legacy application.

Enabling newer applications to speak web services is trivial because they have been developed to allow easy communication with other programs. Modern applications either support web services already or have some sort of XML interface that streamlines creating a web service. For legacy applications or older enterprise applications, the trick is to use the ESA platform to translate the web services XML messages into a format that is native to the legacy application.

Because PCAs and the ESA platform are provided by a vendor, much of the work of creating integration plumbing is done in advance and supported as a product feature. So when a PCA arrives and is installed, the ESA platform starts talking to the existing application in a new way but the current users of the system go right on using the existing application as they have always done, avoiding the need for retraining that frequently drives up the cost of new applications. New functionality is contained in the PCA instead of in the existing applications. Keeping systems intact as much as possible circumvents creating a large amount of custom code or having to abandon the investment in enterprise applications, many of which may have been recently retooled for Y2K.

It is true, however, that the other approaches mentioned in Chapter 2 can leave existing applications undisturbed. Starting over, of course, does not, because it requires a migration from the existing application to a new one. Using an existing system as the foundation for a do-it-yourself ESA platform or creating new composite applications through custom development and integration might also allow existing applications to be undisturbed. The custom development required, however, would not leave the budget undisturbed—the cost of creating custom code and maintaining it are considerable.

Increased value is unlocked from investment in current systems

Once a PCA is in place, a wonderful thing happens: new users get the benefit of information and functionality from existing applications without knowing it. The PCA interface, usually easier to use than the existing application and more tailored to a specific business process, increases the probability that the information will provide value to the company. The PCA also combines information across existing systems in ways that were not possible before, which may provide additional benefits.

This benefit comes through putting the information from the existing systems in a convenient and accessible form and integrating it tightly into the process. In a PCA under development for automating plant maintenance, the application brings together all of the diagnostic information at the time a problem is reported. This information allows the right team of people to be assembled with the required replacement parts so that problems with the assembly line can be fixed in one maintenance visit rather than requiring one visit for diagnosis and another for repair.

PCAs Increase Strategic Efficiency

Strategic efficiency means making the right decisions faster, implementing those decisions faster, and increasing the range of what a company can affordably do. A key element of strategic efficiency is opening the door to new options by lowering the cost of change.

A PCA that supports a strategic process usually automates what can be predicted about such processes, but then leaves plenty of flexibility for the unpredictable events that may change the standard workflow.

PCAs promote efficiency by assembling a comprehensive, unified view of the information in existing systems. PCAs also gather unstructured information like documents, email, and memos, information that is typically isolated in PCs and laptops around the company. By providing access to all this information as well as the ability to add to or modify the information store, PCAs enable a new level of strategic collaboration.

Consider, for example, a mergers and acquisitions (M&A) PCA currently under development. This PCA takes the information that was assembled during the evaluation phase prior to the M&A and allows it to be used during the implementation. This allows the team executing on the merger to understand more thoroughly why the transaction took place and what specific operational efficiencies must be achieved to reach the goals that motivated the deal. Research on the target company's structure, its markets, and products can also be passed on to the implementation team. Although this sounds like a trivial accomplishment, in practice, implementation teams rarely have access to the rationale or research motivating an M&A deal.

PCAs lower the cost of assembling information for decision-making

The process of making a significant decision generally starts by assembling all relevant information. If that information is locked in spreadsheets, end-user project management systems, or enterprise applications that are not easily accessible, then assembling the information may be costly, slow, or incomplete.

One of the core features of PCAs is a unified repository that spans all relevant information from existing systems. This information may be assembled from enterprise applications or from end-user applications like project management tools. The information is normalized in the PCA so a consistent, unified view is provided.

When it comes time to make a decision, information can be accessed immediately, rather than waiting for lengthy rollups. This makes it more likely that decisions can be made as quickly as possible based on the best information available.

The PCA also creates an institutional memory. The basis for decisions, both structured and unstructured information, is captured and does not have to be reassembled for use by teams later in the process.

PCAs allow for more flexible decision-making and project approvals. For example, in some companies, project approvals occur at the beginning of the year so that all of the information can be assembled once a year and projects can be compared as a portfolio on the same basis. With a PCA for project management, the project information flows out of the end-user project management applications like Microsoft Project into the PCA, where it is combined with financial information and other relevant data from existing systems. In this way, the view of all projects as a portfolio is always current and roll-up delays are eliminated. Project approval can occur on a rolling basis instead of just once a year.

PCAs reduce the cost of changing strategy

From the perspective of a PCA, the entire world is set of objects, services, and processes provided by the ESA platform. The ESA platform encapsulates the underlying application so it does not matter where a service is running. It could be on the same machine as the PCA, or it could be on another server in the data center, or it could be across the world.

Because PCAs are separated from the underlying implementation of the fundamental building blocks, a new deployment model becomes possible. The existing applications that form the foundation of a PCA can also be supplemented by services provided by partners or third-party vendors. This flexibility also allows a company to contemplate selling the ability to execute business processes as a service offering to other companies. In this way, IT departments could generate revenue, an unusual achievement.

Changing a strategy and the underlying systems that support it becomes easier and less costly. If a supplier's systems have been integrated into a company's systems, replacing that supplier with another becomes a matter of creating a new adapter for the ESA platform. The encapsulation of the integration with a supplier minimizes the cost of change and prevents inappropriate and expensive integration that increases the cost of change.

PCAs lower the cost of supporting partner relationships

If partner relationships are encapsulated behind a web service, then switching partners is a matter of pointing to a new web service. Switching web services vendors is equally trivial. Of course, differences in data between suppliers might require some translation or mapping, but the costs are reduced to the absolute minimum and the costs of changing can be much more accurately predicted in advance.

This flexibility has profound implications beyond lowering the cost of change. As we will discuss later, the emergence of PCAs will set off a new rush to set standards for components to be used by PCAs. The role of the CIO itself may change to that of a service broker, a person who chooses how to provision applications from a selection of internal and vendor- or partner-provided services.

Naturally, the costs of deeper integration between partners drops further if both are running the same enterprise applications. Web services makes sending messages back and forth easy and cheap. But having the same systems means that the information can be moved in and out of CRM, ERP, or e-commerce systems without transformation because the information is semantically consistent: the formats and data for items like purchase orders or invoices are identical and mean the same thing to the applications on both sides of the conversation. Only large changes that might occur between two versions of an enterprise application will require any transformation or mapping.

PCAs Enable Flexible Cross-Functional Optimization

In the last section, we argued that support of flexible cross-functional optimization was an important goal for the IT department. But how can these processes be supported? What must the systems do? The timeframe for developing cross-functional processes is generally longer than most processes currently supported by systems. The processes are also more strategic, like defining a product, than tactical, like placing an order. These processes involve the entire enterprise, not just a part. All of these characteristics mean that cross-functional processes are less rigidly defined than tactical processes. They change based on fluctuating priorities as issues are discovered and challenges emerge.

PCAs have a comprehensive perspective

The power of PCAs to unify information from a heterogeneous collection of existing applications provides a comprehensive view that is essential to supporting cross-functional processes. In project management, for example, it is not enough to have end-user tools like Microsoft Project all rolled up into a master schedule. Financial information about the budget for those projects is required as well as information about the cost of resources and the skills of people assigned to a project. PCAs do not produce such unified repositories as a by-product. They do so as a first principle.

PCAs capture all relevant information

Strategy is messy. There is no way to anticipate all of the information that will be important to a strategic cross-functional process. Tactical cross-functional processes frequently need to handle exceptions that require assembly of an unpredictable set of information. PCAs are built to extract structured information from existing systems and also to link in unstructured information such as research documents, email, and memos that are produced at various points in a cross-functional process. The ESA platform allows this linkage by exposing both structured and unstructured information in the same development environment. Documents are also kept together and linked to the steps in the process they were associated with, instead of being scattered across a shared file system or locked inaccessibly in someone's laptop. PCAs have functionality for searching and categorizing documents built in. This helps current and future participants in cross-functional processes understand what happened at each step of the process and quickly get up to speed.

PCAs allow for flexible automation of processes

Cross-functional processes range in their scope. They can be precisely defined, like the process of assembling a quote in a sales process. Or they can be extremely vague and meandering, like collecting a set of new ideas at the start of a product definition cycle, which may involve brainstorming sessions and solicitations of ideas from employees and customers. The configurable business process capability of PCAs along with the ability to capture unstructured information allows cross-functional processes to be as automated as they need to be, without losing track of vital information.

Configurable processes are at the heart of the ESA platform. In this way, it productizes the integration of the underlying systems and provides the PCA developer with a homogenous model of the enterprise and the ability to define processes on top of that model.

For strategic processes, partial automation and definition of the process may achieve tremendous results. In mergers and acquisitions, for example, McKinsey & Co. estimated in a recent study that 30 percent of the benefit comes from combining procurement. If a cross-functional M&A process just implements a workflow for improving procurement and increases the speed of that implementation, that will make a large impact without totally automating the process of combining procurement.

PCAs support action

PCAs are not just a dashboard; they are a steering wheel as well. They go beyond rolling up information into a unified repository. PCAs have functionality of their own, and they can invoke functionality of underlying applications, initiate processes within them, and change and create records. This means that PCAs not only help get ready to do something, they actually can contain an interface that allows for action to be taken. In a project management PCA, this might mean approving a plan and then locking in the resources for the team required. For a quote application it might mean scheduling various production tasks in different systems once an order is confirmed. PCAs require less context-switching from application to application because they bring together all applications relevant to the current task into a single, straightforward interface.

PCAs Reduce Costs

PCAs represent the arrival of products for the next generation of automation for the enterprise. As we have noted, in the productization cycle to date, first custom code extends support for new processes and then eventually that support is productized. With PCAs, products are leading the way because the ESA platform allows it. Web services make existing systems more plastic and configurable. Thirty years of IT development has created layers of applications with tremendous power.

PCAs cost less than custom development

PCAs are cheaper than custom development for the same reason that writing a word processor is more expensive than buying packaged software. Because PCAs are products, development and maintenance costs are leveraged over a larger group of customers. Of course, just buying software does not mean that your business problem has been solved. But if a product's functionality meets your needs, it is certain to be cheaper than building a custom version from scratch.

PCAs reduce the cost of innovation

Because PCAs sit on top of existing systems and are constructed by configuring components and services rather than through custom development with languages like Java and C#, they tend to be cheaper than applications like CRM or SCM. PCAs also have a much narrower scope than the current crop of enterprise applications. A PCA may focus on one clearly defined process, like providing a quote, or on a broader process, like project management, but in doing so it relies heavily on the ESA platform layer. This architecture not only drives costs down by productizing integration and support, but it also makes the applications easier to adapt to support new ideas. Custom composite applications or extensions to PCAs are faster to develop because of the underlying ESA platform.

With their lower costs and narrow scope, PCAs allow companies to be more agile and able to consider a wider field of innovation. When creating new systems using custom development, one can never forget the expense of ongoing support and maintenance, and, as a result, a company must be selective when choosing new projects. But when development and maintenance cost less, a company can experiment more. It can afford to have new ideas again as break-even points drop. Smaller projects are also easier to manage and mean less time is wasted in a lengthy RFP process. For a $100,000 project, a manager might be able to approve the expense without sign off from anyone else. For a $1 million project, the CFO is going to want to see a lengthy justification.

PCAs productize boutique expertise

Lowering the cost of development and enabling applications to plug into existing systems should enable smaller firms to bring their ideas and products to a larger marketplace. Boutique consultants who are experts in a certain functional area of an industry will be able to afford to create and maintain products as PCAs in a way that would have been too expensive in the previous architecture. They will be able to leverage the product support infrastructure of the ESA platform vendors.

This results in a win-win situation for consultants and corporate IT. Consultants are able to create focused products and offer them to a larger group of customers. The resulting products are cheaper than hiring these consultants for custom application development. Further, productization pushes ongoing support and documentation costs back to the consultant, resulting in a better product for the customer base. If this were purely one-to-one consulting, companies would pay for the initial development and then again for any upgrades or support they need in the future. Consultants win by broadening their customer base and using their skills effectively to create targeted solutions.

PCAs Manage Change

Because PCAs do not disrupt current applications but bring the functionality of those applications to a broader audience, they can play an important role in helping an organization manage and promote change. PCAs allow for starting small and proving the benefits of an application before asking the entire company to change behavior. PCAs also allow more people to be brought into processes like product development.

PCAs offer a framework for incremental improvement

PCAs provide a way to improve an infrastructure gradually without having to risk traumatic changes. ESA, the architecture of PCAs, is a big step forward, but it is a step that can be taken without disrupting current operations and applications. PCAs are not expensive and can provide significant value with a small footprint. Once one PCA is in place, other PCAs can be easily added because the ESA platform is already installed. The structure of PCAs lends itself to low-cost adaptation to accommodate new partnerships, laws, or strategies, features that are not generally available from previous generations of applications.

If a company learns to use the ESA platform for development, the architecture can be used to create starter versions of applications that may be missing. For example, if a call center application is needed, a simple call center interface can be created quickly using ESA components that allow the requirements to be vetted and verified. Some of these mini-CRM or mini-call center applications may be offered by PCA vendors. Later, if a full-blown call center application is needed, the requirements will be better understood and the risk of implementation would be lowered. A fully featured call center application could gradually replace or supplement the starter system.

PCAS can help promote cultural change

PCAs help reduce the trade-off between the cost of learning a new technology and the benefit provided. PCAs are narrow in scope, easy to learn, and offer a focused, immediate benefit to users. Often, the rewards of learning a new technology take time to reap. For example, in the early days of the Internet, collaborative filtering applications offered recommendations based on how an individual's preferences compared to community preferences. The trouble with many systems was the requirement to enter a lengthy description before any recommendations could be produced. At Amazon, however, if someone bought or expressed interest in a book, boom, there was a recommendation.

PCAs offer rapid benefits to users because they are tailored to a business context. Consider the product definition process, an area described in more depth in Chapter 6. PCAs make it possible for more people to generate ideas for new products. Less than 10 percent of new product ideas come from the Research and Development department. The propensity of people to submit product ideas rises rapidly if they can see what happened to the idea once submitted. Was the idea evaluated? Was it approved, tabled, or were there any comments or questions? PCAs allow more people to participate and get a high-level view of the process as a whole. A product definition PCA that enables broad participation allows various groups to identify objections early in the process. If legal, quality management and supplier management staff can weigh in on product ideas from the outset, problems may be identified that can significantly affect the final evaluation of a product idea. If people know their participation matters, their use of applications can rise dramatically. Changing behavior in this manner can have the effect of increasing the average level of participation in important processes for the enterprise. In this way, PCAs can increase the flow of information to and from all parts of the enterprise.

The Case Against PCAs

So far we have focused on the promise of PCAs and have surveyed important arguments in their favor. In the rest of the chapter, we look down on PCAs from a great height. With a skeptic's eye, we search for possible implementation problems. Our goal is to raise the right questions so that anyone seriously considering PCAs can gain a sophisticated understanding of the surrounding issues and better understand how PCAs may fit into their particular circumstances.

Many of these objections apply more or less to all technology and represent a standard list of questions to ask about enterprise software. Some are contradictory: one objection is that web services can replace PCAs; another is that web services are not ready. Both cannot be true, but both may come up when evaluating PCAs. Some of the objections are specific to PCAs or may be answerable only with respect to a specific implementation of a PCA or an ESA platform. While an analysis of specific products is beyond the scope of this book, the reader will at least know what questions to ask when speaking with a potential vendor.

PCAs Are Not Needed

With every new generation of technology, some skeptics refuse to admit that anything new could help or is needed at all. Here's what we expect such people to say about PCAs.

Existing technology is not fully optimized

"PCAs sure do sound great," the skeptics may say. "Maybe we'll consider them in five years or so, after we have finished implementing all of the new stuff that is just out of the box. We have CRM systems to tune up, and SCM systems to deploy. Once we get those right, the PCAs that sit on top will be all the more powerful, if they are needed at all."

The strongest part of this objection may be the underlying implication that the IT departments in most enterprises are so overbooked that there is no room for any new projects. It is certainly true in most companies that existing technology is not fully exploited, but that does not mean that a more complete implementation is the only choice or the ideal solution set for a company. To accept this argument, one has to believe that the current power and configurability of existing applications will solve a company's problems in the next few years and that preparing for a more flexible architecture will not provide significant strategic advantages.

The second thread of this argument is an attack on the whole notion of cross-functional optimization and automating strategic processes. Existing applications are only going to provide cross-functional support through integration of applications using current technology, which means expensive custom code, the development of which will be fully funded by the company. For cross-functional processes, users will switch from one application to the next, certainly not the most efficient or consistent approach. Strategic processes will not have additional support in this scenario, given their need for a comprehensive view of enterprise information, which again would come only through custom development or integration projects.

This objection is likely to make sense in the rare industry where technology does not play a decisive role in helping a company compete or where strategic efficiency is not important.

A data warehouse is enough

"Can't we get most of the benefits by just grabbing all our data and putting it into the data warehouse, and leave the existing systems alone?" the skeptics will say. "After all, most access to data is read-only anyway, and real-time data is seldom needed. Loading the data warehouse on a nightly basis will be fine for most purposes."

The problem with the data warehouse approach is that it sees all underlying systems as read-only data repositories and ignores the importance of coordinating and integrating the functionality of those systems. Suppose data for an order or a project task is in a data warehouse. During the process of reviewing that data it turns out that something must be changed. Now, the users must go to the existing system, find the record in question, and then make the change there. This context-switching between applications reduces the likelihood that information will be updated.

PCAs can invoke updates to existing systems. PCAs are a way to take action, not just find out what's wrong. Having direct access to the application means that the latest data is available, not just the data from the last refresh of the data warehouse.

The other problem with the data warehouse view is that it ignores the importance of services as an emerging way to integrate functionality within the enterprise and across company boundaries. New applications and partners expect web services, which means the data warehouse approach will need to support web services at some point.

The data warehouse approach also fails to support processes that span existing systems, require collaboration, and demand integration of unstructured data.

End-user tools can do the job

"We do not need a new layer of technology to help us collaborate," the skeptics will say. "We have collaborative technology in place—it's called our end-user applications suite. Email, shared directories, spreadsheets, word processors, presentation programs: This is what everyone in the company knows and these applications are plenty powerful to get the job done. In fact, a huge amount of functionality in these programs goes unutilized. PCAs are overthinking the problem."

It is true that end-user applications are powerful. Who would like to live without email? But then, who has not struggled to find the latest copy among the 10 attachments to a recent spate of emails? How many times have decisions been made without information that everyone knows is present in existing systems but is too difficult to get? How often do we wonder why we made a particular decision and realize that the information on which the decision was based was not captured in an organized fashion? How much work is wasted reassembling information?

It is true that better work habits or more disciplined use of end-user tools would solve some of these problems. But how likely is it that behavior will change unless it is embedded in an interface? PCAs offer an interface tailored to a collaborative process with information provided and captured at the time it is created and distributed. Increasing usability, the ease with which information can be captured and subsequently accessed, will increase the value of the information.

The other weakness of an end-user tools approach is the absence of information from existing systems. End-user tools cannot reliably track relationships between information in distributed repositories. While most existing systems allow data to be exported to end-user applications, consolidating that information into a normalized, unified form is typically done manually in an ad hoc spreadsheet. In PCAs, data unification components do this work so that information is ready when needed.

The strongest part of this argument is the observation that everyone in the company knows how to use end-user tools. That is true and should be leveraged. PCAs that integrate tightly and seamlessly with end-user tools will likely be the most successful.

Web services development tools are adequate

"I went to a conference last week and I saw 25 web services vendors, all of whom said they could deliver exactly what PCAs promise," the skeptics will say. "We spent five years creating an ERP system as a master repository for all of our corporate information, and we have plenty of content management, CRM, and EAI technology. We even have a portal. If PCAs can use web services to bring all this together, well, so can we."

This argument is not wrong, just expensive. Fundamentally, this objection comes down to build rather than buy. It is the not-invented-here syndrome taken to the extreme. Yes, the next generation of web services and application server development tools may be able to build what PCAs promise. Custom development can do anything. It can support cross-functional and strategic processes and use existing systems as services. In addition, a host of vendors stand ready to provide additional plumbing like data unification systems, messaging, and XML unification systems. Web services orchestration and business process management tools can also help.

The question is one of economics: who is going to pay for all these components? Not only must this software be licensed, but the components must be integrated with each other and with existing systems. PCAs provide the whole solution as licensed software, not as custom development. Underneath the PCA, the ESA platform technology provides the functionality of most of the individual products mentioned above, integrated and most likely embedded in the license cost for the PCA. Some of the vendors who provide single pieces of the ESA platform puzzle will no doubt find themselves either integrated in the ESA platforms or purchased by the ESA platform vendors. (We discuss the effect that PCAs and ESA will have on ISVs, systems integrators, and suite vendors in Chapter 5.) ESA platform vendors will no doubt offer the ESA platform as a development tool, but PCAs will lead the way because companies don't want to buy platforms any more—they want solutions that provide business value.

The real threat to PCAs behind this argument concerns the focused productization that PCAs must deliver. To succeed, PCAs must do two things well. The first challenge will be to provide a product that completely solves a specific problem. PCAs must be more narrowly focused than an application with a broad set of functionality like CRM. The second challenge is that the specific functionality must be configurable and extensible in order adapt to surrounding systems and processes in a company without extensive custom development. The more that a PCA project feels like custom development, the less likely that PCA will succeed in the marketplace. Achieving configurability and extensibility while maintaining the identity of the PCA as a product that can be maintained and upgraded is the key challenge for PCA vendors.

PCA claims are broad and unrealistic

"Look, the whole idea of PCAs is just great. But how is it different from a portal? How is it different from reengineering? Buy low, sell high is also a good idea," the skeptics will say. "All this talk of cross-functional optimization and strategic efficiency is just some marketing drivel. The benefits claimed for PCAs are benefits that cannot be derived from technology. They come from good management crafting a vision and leading the company to adopt it. PCAs can't provide that leadership, yet they claim to."

In many ways, this particular objection is the reason this book was written. The most profound response, and the only one that will silence skeptics, will be PCAs in production at hundreds of companies making a significant contribution. The proof is in the pudding, not in the claim.

But even at this early stage, we believe we have a compelling case for PCAs. For example, we show the mechanisms PCAs will use to support cross-functional optimization and strategic efficiency in Chapter 6. These specific examples show how cross-functional optimization and strategic efficiency provide tangible benefits. We find the weight of evidence overwhelming. It may be possible to argue against some of these examples, but not all of them.

It is hard to disagree with the proposition that good management is more important than technology. However, we believe that PCAs offer advanced support for good management. As we explained in Chapter 2, IT support has expanded to automate more and more business process and we propose that the next frontier will be cross-functional and strategic processes. Furthermore, we argue that these applications will have patterns, which will determine the scope and functionality of the first generation of PCA products. The first PCA vendors are already surprised at how quickly their products are applicable across vertical markets. We respectfully disagree with this sort of objection and point to the weight of evidence we present as a response.

PCAs Will Be Disruptive

In theory, there is no difference between theory and practice. But in practice there is. So it will be with PCAs. The following objections point to high-risk areas, battlegrounds on which PCAs will succeed or fail.

Preparing the existing systems will be problematic and expensive

"I like the idea of PCAs, but isn't the hard work in preparing the existing applications to be services, and won't that be expensive?" the skeptics will say. "Legacy applications, in particular, are rife with duplicate data and are ill-equipped to provide services to other applications, especially with respect to issues like supporting two-phase commit transactions. This is much more work than you are suggesting."

This objection points right to the highest risk and highest cost area for most companies: legacy applications. These programs were developed a long time ago, frequently with older technology or with a brain trust that has left the company. For these reasons, they are expensive to maintain and change. Legacy applications are also known by another name: applications that work. It is not unusual for a legacy application that has stood the test of time to provide unique, strategically valuable functionality. If it doesn't, then it should be a candidate for replacement.

The cost of preparing a legacy application to be a service varies. The worst case is a batch system, which may be difficult to turn into a service. There are few ways of getting around this barrier except to focus on the database. On the other hand, mainframe systems are not nearly a lost cause given how much work has gone into providing gateways between Java and mature applications like CICS.

But many of the problems with legacy applications are problems whether or not they provide web services to PCAs. If there is bad data in an application, if it is expensive to maintain, if it cannot integrate with other systems, these are problems in any event. At least adapting such an application to a PCA allows the legacy application to be extended at a reasonable cost by putting functionality in the PCA layer. It also provides a path for gradual migration. Exposing a legacy application as a service to a PCA also makes the functionality available to a wider user base that may benefit from it.

The question is, What is good enough for the legacy application to play a productive role as a service in a PCA? In some cases, read-only access on the database level may be good enough. The lack of two-phase commit transactions can be overcome with alternative approaches such as compensating transactions.

The enterprise computing environment is not ready for PCAs

"It is great to talk about the easy part of PCAs, the job of aggregating functionality," the skeptics will say. "But what about the looming issues of single sign-on, web services support for transactions and guaranteed messaging, and incompatibilities among operating systems. These issues will place a firm roadblock in the way of the PCA vision."

There is no doubt that the enterprise computing environment is more mature in theory than in practice. One of the reasons that web services have so quickly overcome EAI is that EAI implementations have been slow to proceed. Issues like single sign-on are also solved at many corporations on the drawing board but not in practice. We will examine the readiness of web services in a later objection.

But we reject the notion that this is a roadblock. The lack of a fully mature enterprise environment has not been a roadblock to integration when it mattered. It has caused various security problems to be addressed on an ad-hoc basis, which has increased maintenance and administration, but maintained security.

This objection echoes those covered earlier that point out that existing technology is not fully optimized and that underlying systems are not yet fully deployed. CRM systems have often taken two or three tries to get right. SCM systems may only be partially implemented.

Our response boils down to this: waiting for the enterprise environment to be perfect or waiting for existing applications to be completely deployed and optimized delays the benefits of PCAs, which are significant in their own right and amplify the investment in the existing application stack. The vision of a perfect system should not keep us from improving the current imperfect one.

PCAs will not be meaningfully decoupled from existing systems

"You know, the concept of the service-oriented architecture is not new, nor is it easy to get right," the skeptics will say. "The whole notion of refactoring exists because it is difficult to encapsulate functionality without experience in production to see what works properly. The PCAs offered will probably be much more sensitive to changes in underlying applications than expected and will not work properly. They may also have enormous maintenance costs."

This objection rings true for anyone who has participated in the launch of a new system and the subsequent struggle to adapt it to newly discovered requirements while in production. The fact is that the interaction between the business process and a supporting system may not be thoroughly understood until a year or so goes by and the third major release has been installed.

But look at the enterprise environment. The rookie year has passed for almost every application. Spending for new systems has been turned off for some time. There is significant experience in production for the most important applications and either the vendors or the IT staff knows what information is most important and needs to be shared. When versions change, there has to be a well-designed process for migrating important data, which in any event is generally stable. APIs do change, which means that the code that supports the web services layer may have to be modified when an existing system is upgraded, but for PCAs that cost is borne by the vendor, which provides connectors for the existing applications.

The other factor working in favor of PCAs is that they are created from components that are more mature and configurable than previous generations of technology. Coding has not been eliminated, nor will it be, but PCAs will probably be constructed more through semantic modeling than through more traditional approaches.

Finally, application suite vendors who sell ERP, CRM, SCM and a variety of other enterprise applications will likely be ESA platform vendors. These vendors will integrate their ESA platform with their own technology, which reduces the cost and risk of getting things right. And even if they don't create an ESA platform, all enterprise application vendors are focusing on making their technology as friendly as possible to web services.

PCAs Will Not Work

Here we take a look at the direct attacks on PCAs, objections relating to the ability of PCAs to work as promised. These objections are speculative in that they can only be answered with reference to a particular PCA or ESA implementation, and these are just entering the marketplace. But for anyone who is seriously considering using PCAs in the future, it is worth considering the worst case.

ESA platform vendors will never attract developers

"Both ISVs (independent software vendors) and IT departments are wary of making a huge investment in a new platform," the skeptics will say. "And that's just what the ESA platform vendors are asking them to do. If the developer community isn't won over, then the PCAs with the most value, those that come from ISVs and systems integrators with the most knowledge about specific business processes, will never arrive. PCAs will extend a specific vendor's application, but will not be a viable new application platform across all vendors."

This argument points to a crucial issue for ESA platform vendors both in strategy and execution. The most important part of this objection is the notion that third-party developers in the ISV and systems integrator community may never adopt the ESA platform. At first glance, perhaps it is a stretch to think of a group of developers switching *en masse* to a new way of developing applications. But if we look a little deeper into the forces at play, we will see some powerful motivations for potential developers of PCAs. (We delve much deeper into the issues regarding ISVs, systems integrators, and suite vendors in Chapter 5.)

This objection represents a misunderstanding of why developers choose a platform. It is not primarily because of its technical benefits, but as a way to create salable software. ESA platform vendors will open up large new markets to ISVs and systems integrators because software developed on their platform will be lowest risk when it is installed in their customer base. For an ISV with a clever new way to optimize a business process, it will be a lot easier to sell components that sit on top of a layer approved by the major vendor for a customer than to elbow their way in with a new layer of software that must be integrated.

That said, the development environment offered by ESA platform vendors will have to be not just good, but great. Current development environments from operating system and application vendors are in their fifth, six, or seventh versions, and they make life easy for developers. The bar is high. But more importantly, the ESA platform allows a suite vendor to go after customers where other suite vendors are entrenched. The ability to integrate across heterogeneous existing systems makes the whole application suite market fair game to any of the vendors. The vendors with the best development environment will have the best PCAs to offer and that will be a powerful force driving adoption. An open question is whether suite vendors, who have not been historically strong in the development tools market, will produce ESA development platforms that make developers happy.

ESA platform vendors will not be open

"So, for PCAs to catch on, suite vendors, who for years have tried to be the comprehensive vendor to their customer base, will now embrace a third-party development community," the skeptics will say. "Yeah, right. First of all, they don't know how to build and support relationships, and second of all, they are used to keeping all of the pie for themselves."

This objection is correct in its observation that suite vendors will have a new game to play when they become ESA platform vendors. The suite vendors who build relationships the most effectively will produce the most PCAs, which will spur the adoption of the vendor's ESA platform.

But this objection is dead wrong in its claim that suite vendors are not used to sharing revenue or ownership of the customer. Almost every major engagement for suite vendors involves a huge contract for a systems integrator. A complex integration with an enterprise application vendor of a CRM or SCM system is also a common feature of an implementation of a suite vendor's products.

Standards will be needed but slow to come

"The best thing about standards is that there are so many to choose from," the skeptics will say. "With PCAs, a whole new wave of standards wars will develop as the vendors for domains like knowledge management, content management, and business process management all present their versions of how to expose their functionality within the ESA platform. On top of that, each vendor will have proprietary functionality that will be available as extensions and the PCA developers will have to decide which if any of these to support, which may result in an unsupportable mess."

It is indeed true that standards will develop for the component parts of the ESA platform and for the interfaces to underlying services.

And in some cases, not all interested parties will agree and competing standards will exist. But with web services and XML firmly in place as the way to express standards, the cost of supporting competing standards has dropped dramatically. If a vendor loses a standards war, the cost of supporting the winning standard is small. The technology community's appetite for standards wars has also abated to some degree. It is becoming more common for large vendors, even those who are able to fight at length, to agree to support a standard as a service to the user community.

It is also possible to argue that standards are a following indicator, not an enabling force. According to this view, standards will become important after a PCA has established a need for a particular service.

Web services are not ready for prime time

"Web services are often praised but seldom put into production," the skeptics will say. "And when they are, it is safely within the bounds of an enterprise with no messy issues like security or guaranteed delivery to deal with. If you look at the number of issues outstanding within the web services standards process and the progress being made, I wouldn't bet on web services being well-defined for several years."

It is clearly true that web services hype is far ahead of the quality of the standard or the implementations based on it. Basic issues such as the nature of the implementation paradigm (XML Messages or RPC-style) are still being contested and there are gaping holes in the standard. But the most important part of the game is won: everyone believes that web services can make their applications communicate with one another. It is like the beginning of the Web's expansion in the mid-1990s, when countless hours were wasted as people waited for web pages to creep across their 28.8K modems. Nobody would have predicted that people would put up with such a low quality of service, but indeed they did, and the problems were rapidly solved.

The same will happen for web services. Vendors will make do and code around gaps in the standard, which in turn influences the standard. Because web services and XML are so flexible, it will be cheap and easy to adopt new layers as they emerge.

PCAs are promising too much flexibility

"So PCAs will be more configured than programmed, more modeled than coded. Isn't this the death of programming spiel all over again?" the skeptics will say. "Coding is alive and well, and PCAs won't change that. The configurability, which will speed their deployment, will never arrive. PCAs will just be applications that use services as inputs with no special advantages."

Well, one response to this objection is that the worst case still leaves plenty of room for PCAs to provide tremendous value to end users, who don't give a hoot how an application was developed. And it is wise to be skeptical of the claims of the death of programming, but PCAs really don't claim that. They claim a reduction in programming because the fundamental act for PCAs will be gluing together services. The plumbing to do this is not new and it began development as methods to map XML messages to one another and add functionality to certain tags. Scripting is also far more mature than it has ever been in its application to enterprise software. Even though programming will play an important role in PCAs, configuration, modeling, and mapping will play even larger roles in these products than in any previous generation of software.

Regardless of how many arguments for or against PCAs may have resonated with you, after reading this discussion, we think it is hard to ignore PCAs as a potential direction for IT. The logical next step in evaluating PCAs is to look underneath the surface at their deeper structure, which is the purpose of the next chapter.

4

Anatomy of a PCA

PCAs are a concept, an idea for how applications should be built in an affordable, flexible way that is likely to produce significant value for a business. So far we have described PCAs as a relatively large black box, one that, as we have mentioned, is the application delivery vehicle for Enterprise Services Architecture. In this part of the book we open up that box and look inside to see how it is constructed.

What, though, is the anatomy of a concept? For a specific implementation of a software product, the anatomy is the architecture, the way that the architects and developers have organized the software into several different components, defined what each component does, and then connected the components to solve the problem at hand. For our purposes, the anatomy of the PCA concept includes a tour of its structure and a survey of the questions facing its builders and implementers. We will look at all the work that a PCA must do and then describe in a general way how that work will get done.

PCAs are built on an ESA platform, a software product that implements the ESA vision. One of the most fundamental questions we will address is what kind of building blocks the ESA platform will offer architects of PCAs. Given that at the time of this writing in mid-2003 PCAs are just coming to market and ESA platforms are under development and are appearing in their first versions, we will not be flying blind. Based on interviews with architects of PCAs and ESA platforms, we will be able to delve quite deeply into the most bewildering questions and challenging tradeoffs that arise in implementation.

As we have seen so far, the vision of PCAs and the ESA platform is not for sissies. The fully realized PCA/ESA paradigm achieves incredibly ambitious goals, including pulling together all the loose ends from 30 years of IT history. The questions that arise in creating a unified view of existing enterprise applications and then building the sort of applications that can support rapid innovation go right to the heart of software architecture and integration issues. How can objects from a variety of different underlying applications work together? How should the ESA platform be layered to provide application-specific services as well as generic services? If an ESA platform will not spring forth fully formed, like Athena from the head of Zeus, in what stages will such a platform be developed? How can all of the unified components be glued together to form an application?

In answering these questions, we will continue to make the case for PCAs and show how they offer the way forward for the next generation of enterprise computing.

PCAs and the ESA Platform

A PCA is not just a product that does its job by using functionality from other applications. A PCA is actually part of the Enterprise Services Architecture. The vision for ESA, which we will describe in the next portion of this chapter, is one that reorganizes the infrastructure for developing, packaging, and delivering applications. PCAs are the container and delivery vehicle for the application functionality in ESA. Just as application code is written using the tools provided by an application server, so PCAs are written using the tools provided by the ESA platform.

As Figure 4-1 shows, PCAs are built on top of an ESA platform, which creates a unified model of all current enterprise applications.

At the highest level, we have the PCA, which uses the objects, services, and processes of the ESA platform. The ESA platform pro-

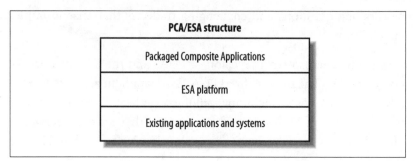

Figure 4-1. Enterprise Services Architecture

vides homogeneous, unified access to the existing applications and systems at an enterprise. The term "Enterprise Services Architecture" incorporates all three of these layers.

One important point that for simplicity's sake we have not stressed so far is that many PCAs can sit on top of one ESA platform and on top of each other, as shown in Figure 4-2. A product definition PCA that collected ideas from employees might be able to provide input to an employee evaluation PCA so that at review time a manager could look at all the suggestions made by an employee.

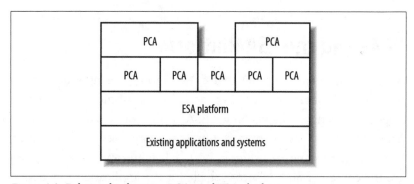

Figure 4-2. Relationship between PCAs and ESA platform

To a user, all of this is really not visible. Users see an application that brings together all of the information they need to do their jobs. But the idea of PCAs as components that use other components, which may in turn use other components, will be a theme in this discussion.

As applications, PCAs bring information together from wherever it resides in the enterprise. They enable flexible automation of processes so that steps in the process can be modified. PCAs also provide integrated collaboration about the information in the application using email, threaded discussions, phone conferences, and so on, and they supplement the existing infrastructure with new applications.

What the users do not see is that, under the covers, PCAs are a whole different animal. To the developer, ESA changes everything. The best way to get at these changes is to start from the goals of ESA and then look at the building blocks it provides.

The ESA Vision

The ESA vision springs out of the current state of enterprise architecture. Significant investment in the latest generation of technology began with the three-tier, client/server ERP systems and radiated out through enterprise applications like CRM, SCM, and others. The single-minded focus of the developers of these applications automated a huge amount of business processes, but in a partitioned way. Each application was bound within the silo of functionality that it owned. Within that silo, the applications were configurable, but processes that crossed the application silos were hard to automate. This sort of cross-silo automation, also referred to as cross-functional automation, can be accomplished through Enterprise Application Integration (EAI) technology, which automates the processes but does so in a way that is expensive and difficult to change.

The ESA vision is to remove all barriers to automation so the developer has the ability to look at any object, any service, any process in the infrastructure and combine them into a new application, adding functionality where it does not exist and reusing it where it does. The ideal is that the process of improving a business is bound by the creativity and imagination of the executive team, not by the inflexibility of the technology infrastructure. The ESA vision has the following characteristics:

- The ESA vision is to create a new, unified infrastructure through leveraging existing enterprise applications and platform components, which is likely a heterogeneous mix of functionality from a variety of vendors with custom legacy applications mixed in. The ESA platform, which will implement the ESA vision as a software product, will provide a unified, homogenized view of the enterprise to empower developers of PCAs and of custom composite applications.

- The ESA vision is to package and productize integration between application silos so that the ESA platform in effect creates a system through which point-to-point connections between applications can be constructed as a matter of low-cost configuration rather than expensive hard-coding.

- The ESA vision is to plan for change and incremental improvement by putting flexible and configurable business processes at the center of PCAs. This feature, along with packaged integration, provides the foundation for rapid cycles of innovation, implementation, evaluation, and improvement.

- The ESA vision sees everything as services, which encapsulate applications and provide a simple gateway to their functionality. In an ideal vision of services, common abstractions or models of the enterprise objects, services, and processes are created that can be glued together not with programming languages but through developer-friendly frameworks that allow configuration through a graphical interface that conceals the complex implementation details of the underlying applications.

- The ESA vision is to connect and integrate the worlds of transactional applications that manipulate structured information, such as records in databases, and the collaborative world of unstructured information, such as email and online chat, so that the developer can provide applications with both sorts of functionality, seamlessly working together.

This, as we have said, is a grand and glorious vision that, if realized even in part, would provide a path for companies to transform the tightly coupled, hard-coded, heterogeneous, expensive-to-integrate, and inflexible infrastructure of today into one that enables change and productivity. The rest of this chapter will explain just what such a platform would look like and how it would be used to build PCAs.

Gross Anatomy of the ESA Platform

Now that the goals of the ESA vision are clear, what then must the ESA platform do to achieve them? What would a unified, homogenous view of the enterprise look like? What sort of magic would turn the existing heterogeneous infrastructure of today into such a pretty picture?

To answer, we will first review some basic facts about the existing infrastructure. Our analysis divides the existing infrastructure into two categories: enterprise applications and platform component systems.

Enterprise applications are generally large software programs aimed at automating a specific collection of business processes. ERP, CRM, and SCM are all enterprise applications.

Platform component systems are large toolboxes of functionality designed to create applications that solve certain specialized problems. Platform component systems include content management systems, knowledge management systems, data warehousing systems, and other such programs. We also include pure general-purpose systems like application servers or EAI systems in our definition of platform component systems.

These definitions are not precise, but software packages generally fit in one category or another. As software products mature, they tend to offer more and more functionality from both categories. A CRM package, for example, starts out as an application for end users, but adds more and more ways to use the application as a toolkit for integration and other purposes. On the other hand, content management systems usually start out as toolkits and then, as they mature, offer vertical applications to solve specific problems.

The Application Stack

The next step in describing the ESA platform is to look at the structure of enterprise applications and platform component systems. We can use the idea of the application stack to look at both types of software through a single lens.

The application stack is the name for the different layers of functionality that exist in most software programs. The layers that we will describe are as follows:

The user interface layer
  The part of the software that controls interaction with the user.

The process layer
  The part of the software that automates business processes, moving end users from one step to another as they complete tasks and coordinating the execution of underlying services.

The services layer
  The part of the software that contains the logic to perform the transformations, calculations, and other required processing to do the work of the application.

The object layer
  The data of the application and the associated service functions that perform basic manipulations on the data and move the data back and forth between the object and persistent storage.

The persistence layer
  Usually a database, although it is possible to store or persist data in many ways, so this more general term is used.

Not all software has every layer of the application stack—a message broker may not have a persistence mechanism or much of a user interface. A CRM application will have functionality at every layer. An application server will usually provide tools to help build functionality at every layer.

To do its job right, the ESA platform must create a two-way street between the PCAs and the enterprise applications and platform component systems whose functionality will be used to do the work. But in order to make sense of the varying levels of functionality in each of the underlying applications, the ESA platform adds a new dimension to our view of application functionality: the degree to which a specific object, service, or process is general purpose.

To understand this, let's first imagine that the application stack was turned on its side as shown in Figure 4-3.

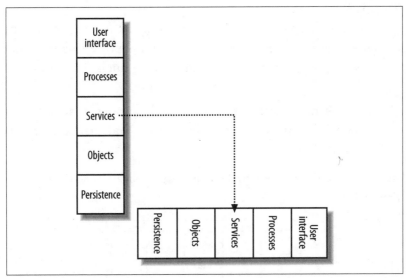

Figure 4-3. ESA application stack

We can imagine all of the functionality of the underlying enterprise applications and platform component systems sorted into these categories. The user interface layer would contain the user interface functions of the application server and of the CRM system. The functionality of these two programs differs immensely. The application server provides generic user interface functionality that can be used to create a user interface for any application; it is not tailored for a particular purpose. The CRM user interface layer typically provides functions to display customer records and contact information at varying levels of detail. Some of the user interface functionality in the CRM application is specific to a particular page or to a task that is performed only once or twice in the application.

To build an ESA platform from a set of heterogeneous components, we add three more layers that are orthogonal to the application stack to help us make sense of the differences in generality between the different types of functionality. The three layers are:

The PCA layer
> Contains application-specific functions that are not shared with another PCA.

The application services layer
> Contains the functionality that is designed for a specific purpose, that is aware of the kind of data it is processing, and that may be shareable among PCAs performing similar functions.

The generic services layer
> Contains functionality that is general purpose, that does not know what sort of data it is processing, and that is shareable across all PCAs.

Adding these three layers makes the ESA application stack look like Figure 4-4.

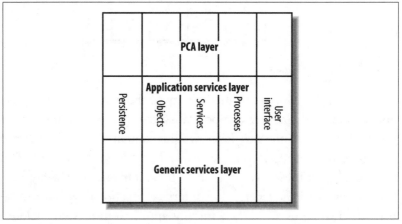

Figure 4-4. ESA application stack layers

The three layers differ from each other on additional dimensions besides application specificity. They also vary in how much functionality is aggregated and how standardized they are.

The generic services layer is atomic. Each service knows little or nothing about the other services at its layer. Each service is a world unto itself that performs a function without using other services at the generic services layer. Opening a file is an example of a generic service.

The application services layer is composite. Application services may use many different generic services to do their job. They may use other application services. They are much bigger and more complex than generic services. Creating all the records needed to keep track of a new account and take an order is an example of an application service.

Generic services are standardized. The data and services offered by the generic services layer do not change from application to application. Generic services are the container for the functionality that has been commoditized.

Application services are customized for a particular purpose. The data used in an application service, like a purchase order, may differ from company to company and application to application. Applications services, however, may be shared by a group of applications that are using the same objects, processes, and services.

The PCA layer sits above the application and generic services layers and is the most unique and specific layer of all. The PCA layer functionality is not shared and is specific to a particular application.

These layers exist in each part of the application stack. Processes can be in the generic, application, or PCA layer, user interface elements can be in any layer, and so on. Figure 4-5 shows these three layers in more detail.

In a way it would be more appropriate to call the generic services layer something like the "generic, atomic, standardized, very shareable layer." In this long form the application layer would be the "application, composite, customized, partially shareable layer." The short names are much nicer.

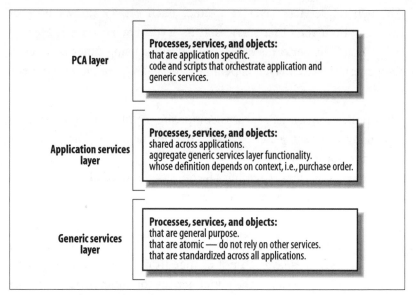

Figure 4-5. Definition of ESA layers

With these three layers, we may partition the functionality of the enterprise applications and platform component systems in a way that allows functionality from the systems to be organized into different categories. Unfortunately, such a nice conceptual framework is not the end of the story. Many questions must be answered to make this categorization into a product that can be used by developers. The way that objects, services, and processes are defined and work together must be specified. The way they are connected at development time and compiled and executed at runtime must be discussed. We now turn to the thorny issues of implementation.

How to Build an ESA Platform

So far we have papered over how all the layers are implemented and connected. Now we will delve into the details. The idea is not to describe a specific implementation; we will stop well short of that. Our goal is to describe the different ways that the ESA platform will likely show its face to PCA developers and eventually to IT departments who will use the ESA platform for custom development.

An ESA platform would be a purely theoretical idea with little chance for implementation were it not for the rich set of technology and infrastructure currently in place in the corporate computing environment.

ESA Building Blocks

A service-oriented architecture is one of the fundamental concepts at the root of the ESA. The crux of this concept is that computing resources can be defined as a set of services that encapsulate the functionality of underlying applications and allow access to that functionality through a relatively simple interface. Properly defined, the services should be able to be loosely coupled; that is, they should be able to be combined over and over again to solve different problems in a way that avoids dependencies between services or unintended side effects from invoking a service. In a service-oriented architecture, the components that offer services should not only be well-behaved providers of services, but they should also consume services in an organized manner. That is, the implementation of services should employ other services where appropriate, resulting in a very flexible architecture that allows services or components to be replaced or improved with as little impact as possible on other elements.

Web services are the modern face of service-oriented architecture. They provide a way to present the services of each of the layers described above. Because so much energy is currently being expended to make enterprise applications and platform component systems speak web services, it is likely that web services will play an important role in creating an ESA platform. Objects, services, and processes at the generic, application, or PCA layers could be exposed as web services so the developers could orchestrate their behavior to build applications.

Web services are not the only game in town, however. In fact, web services, despite their promise, will probably not be a viable choice for a significant number of tasks in an ESA platform because current

implementations lack support for features like transactions, security, and guaranteed delivery. Technologies like EAI systems and
good old APIs have these vital features and they will clearly be
employed in building the ESA platform.

Business process management tools that allow the definition of a
process that can then create, invoke, monitor, and control objects,
services, and other processes will also play a significant role in the
creation of the ESA platform.

Modeling environments that support abstract descriptions of objects
and processes and allow them to be connected to services will also
be important building blocks for ESA platform architects.

Figure 4-6 illustrates how an ESA platform puts all of these moving
parts together.

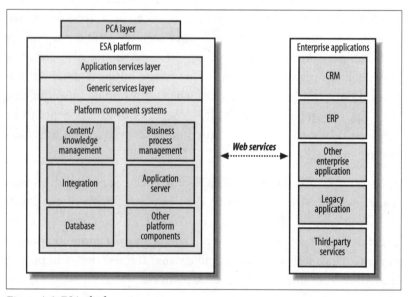

Figure 4-6. ESA platform structure

Constructing an ESA Platform

The process of using these tools to knit together all of the functionality offered by the underlying enterprise applications and platform
component systems into a well-formed, easy-to-use set of objects,

services, and processes remains a complex task, no matter how much help is available.

The first step is to find out how to access the data at the heart of the applications. This data will reside in objects, mostly at the generic and application layers. As part of an ESA platform, an object model for all data contained in the underlying applications will be created. The data in these objects will come from the enterprise applications and may be manipulated by services provided by platform component systems. A basic customer object with only fundamental customer information, based on an industry standard, might be an object in the generic layer. An extended set of customer information for a particular purpose like holding detailed information pertaining to an account could be an object at the application layer. An object with all the information for a customer from all levels including information about how recently the application was accessed might be in the PCA layer.

Services from the underlying applications must be made available in the ESA platform at the appropriate level. It is likely that an application will have one preferred way of providing access to its services, even if many methods are supported. The ESA platform must be able to use all these methods, including EAI, SOAP, web services, and APIs, and then expose the services to developers who will use them to transform and manipulate objects. For example, a content management system's indexing and searching functionality might provide an indexing and searching service for objects.

Processes from applications may be exposed and employed by the ESA platform. It is also likely that PCAs will define new processes. Processes like steps in account creation may be one sort of process that could be exposed in the ESA platform.

Now things start getting exciting. The objects that are created must be able to move data in and out of the underlying applications. They must be able to sort out which data in the object should be populated from which enterprise application and should also be able to save any changes to the correct application. The services at each layer should be able to operate on the objects. Perhaps all services will be able to access all objects, or perhaps certain services will apply only to subtypes of objects. The processes that orchestrate the behavior of the objects and services must also be connected to them in a consistent way.

An ESA platform must create an ecosystem, an environment in which objects, services, and processes that are implemented using services from underlying applications can be created and work together. Figure 4-7 shows how a CRM system might expose its functionality within an ESA platform.

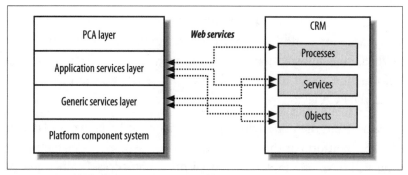

Figure 4-7. ESA platform interaction with enterprise applications

The difference between design time, run time, and operational support makes this picture still more complex. Design time is when a developer is sitting in front of a computer terminal cobbling together objects, services, and processes into a program that performs some useful task. If an ESA platform provides developers with a 20-foot stack of documentation for all of the APIs that need to be connected to make the ESA platform work, it is unlikely that it will be a success. On the other hand, there is a boatload of complexity in the picture of an ESA platform we have just described.

The likely solution to this problem of empowering developers to create applications in an ESA platform will involve some combination of modeling and application generation. The programmer will describe much of the way that objects, services, processes, and user interfaces are supposed to interact using some high-level modeling environment. The program to support the relationships so described will then be generated and the developer will be allowed access to locations for code in which modifications and customizations are allowed to complete the task of building the application.

The application constructed in this manner would have to be highly configurable, so that each time change was required the entire generation and compilation process did not have to be repeated.

Support for the monitoring of the behavior of the application would be provided as it is for all software products, but because the ESA platform is built to use external services, special care will have to be taken to allow operational personnel to determine whether a problem is in a PCA or in the underlying application.

Major challenges await the architects and developers of ESA platforms. One of the most thorny is how to implement support for transactions across the underlying enterprise applications, some of which may not expose transactional support in their APIs. Transactional support means that the beginning of a transaction can be declared and then a series of changes made to a database or other persistence mechanism. If everything goes well, the transaction is committed, and all changes are saved in the database. Otherwise, the transaction is aborted and no changes are made. Transactional support is a crucial requirement of many applications.

The other challenge that will be quite daunting is using enterprise applications, many of which were not created with the idea of being a service to other applications, without causing unintended consequences. It can be difficult to understand the consequences of using an API or service interface to an application. Making sure that the behavior of a service stays the same as the application moves from version to version can also be troublesome.

Creating a unified model of all of the objects, services, and processes will also be quite a task, one that will probably not be completed properly until after several versions of an ESA platform. Because of the challenges and complexity involved, the ESA platform will come to fruition in a series of stages, the nature of which we will describe next.

Stages of ESA Platform Evolution

The stages of the development of the ESA platform are an indication of the order in which the challenges will be met and perhaps of their degree of difficulty. ESA platform vendors bring different assets and strengths to the task, as we will discuss in Chapter 5. These differences mean that each ESA platform vendor may solve the problems described in a different order or perhaps package the ESA platform in a slightly different way. Even if these stages fall short of predicting the future, they will give the reader a framework for analyzing different products.

Regardless of what an ESA platform vendor brings to the table, constructing such a product is a huge undertaking. The variety of layers and problems is so complex that it is unlikely that they will be fully understood until the first versions of the platform have been released and are supporting PCAs in production. It is then that the ESA platform architects and developers will realize how to improve the platform and get it just right. Remember that the first versions of the ESA platform are likely to be used by PCA vendors and close partners of the ESA platform vendor. The good news for the corporate IT developer community is that by the time the ESA platform is released as a general development tool, it will have been tested in production by the PCA vendors. In other words, the complexity and difficulty of the task of creating an ESA platform along with the rush to get the PCAs to market encourages a healthy iterative development cycle.

Data Versus Functionality in Underlying Systems

Before we describe the stages of ESA platform development, we must first examine the differences between how enterprise applications, such as CRM systems, and platform component systems, such as content management systems, will provide their functionality to the

ESA platform. At a high level, the difference is between data and generic functionality. Enterprise applications have data but limited generic functionality. Platform component systems have generic functionality, but no data.

The impact of this on ESA platform architecture is that at first it is likely that the power to manipulate information and objects will come from the platform component systems. Some sort of application server will form the core of most ESA platforms. The application server provides the basic plumbing for creating objects and services and allowing them to interact with each other and with other services from inside and outside the platform. Surrounding this will be the functions of the platform component systems for content management, EAI, business process management, data warehousing, and other generic functionality. In other words, platform component systems are the kings of the generic services layer.

The data in enterprise applications represents a fundamental asset of the corporation. Manipulating this data and moving it around is the reason we have an IT infrastructure. Enterprise applications generally have pretty good mechanisms for importing and exporting data, either through XML interfaces or APIs. However, enterprise applications do not excel at exposing their functionality to manipulate that data. Enterprise applications have mechanisms to find data within the applications or to invoke certain processes or inquire about the results of processes and the like. But much of the valuable features and functions of most enterprise applications is concealed behind their interfaces and APIs. Further, the features that are exposed were not designed to be part of a service-oriented architecture. In some cases, invoking a service in an enterprise application may have unintended consequences. This means that in the early stages of ESA platform development, enterprise applications will be the kings of data, especially at the application services level, and will provide services that can be used at the application services level as they are able.

Inside Versus Outside the ESA Platform

From the PCA point of view, the ESA platform provides objects, services, and processes to allow applications to be created. How those objects, services, and processes are implemented are the ESA platform's problem. The ESA platform, however, clearly demarcates what is inside the platform and what is outside the platform. One of the key value propositions of the ESA platform is to package and productize the integration that brings enterprise applications and platform components inside the ESA platform so they can be used to construct PCAs.

In a world of services, however, what do we mean by inside and outside the platform? Don't services, especially web services, obscure exactly where a service is provided? This line of questioning opens up a discussion about how to define what "inside" the ESA platform could mean. We are going to keep things simple and define the functionality inside the ESA platform as functionality that exists within the ecosystem of the platform and does not require an adapter to intermediate. For the first versions of ESA platforms, this means that platform component systems are the only systems that will be inside the platform, and enterprise applications will be outside the platform because they will require adapters.

An example will help make this clearer. Let's say that a PCA creates an object containing the text of an email from a collaboration system. The PCA wants to make sure that this text is indexed so that it can later show up in a search engine. If the text search system is inside the ESA platform, then the object will be able to invoke the search service directly without an adapter. Depending on how the ecosystem works, this invocation may take the form of an API call or a web services invocation, but on the other side of the invocation sits the functionality, not an adapter. If the search engine were outside the ESA platform, then some sort of adapter would manage the conversation between the object and the text search system.

What do adapters do? They manage the conversation between applications. They transform data from one format to another. They mediate between different languages or methods of communication. They handle the messy details.

The more functionality that is outside the system, the more adapters will be needed. Most enterprise applications will be outside the system in the first generation of ESA platforms. As the ESA platform matures, enterprise applications will likely be written inside the ESA platform. Legacy applications will always be outside the ESA platform. The core platform component systems of the ESA platform will be inside the platform, but there is plenty of room for specialized or alternative platform component systems to be outside the platform.

The idea that it may be possible to choose a platform component system that is inside or outside the ESA platform brings up an interesting challenge for ESA platform developers. If the ecosystem allows a platform component system like content management to be implemented either with the functionality that is inside the system or outside the system, that means that the functionality of the content management system is encapsulated in some sort of abstraction so that the objects can invoke the functionality without knowing where it is being implemented. If it is inside the system, then the function is invoked directly. If outside, then it is invoked through an adapter. This brings up some interesting questions. If web services are used as the abstraction mechanism, will performance be hurt by all of the XML conversion? If the functionality is abstracted so that any number of content management systems could be used, then how will that abstraction be defined? Will there be a standard interface? If there are extensions beyond the scope of the standard interface, how will that functionality be made available?

Questions like this abound in the design and construction of the ESA platform, which is why it is sensible for the platform to be developed in stages.

Stage 1: Basic Functionality and Adapters

The initial stage of ESA platform construction involves getting all of the objects, services, and processes from the enterprise applications and platform component systems into one ecosystem that works well enough to support the creation of the first crop of PCAs.

A Stage 1 ESA platform will likely have adapters to access data in all of the major enterprise applications. The adapters will allow data to move in and out of objects at the generic and application services layers. Some key functionality, mostly involved with searching for data, will be exposed from the enterprise applications as services. In an HR application, for example, the ability to search for an expert in a certain area would be made available as a service, probably in the application services layer.

With some sort of application server at the center, the most important platform component systems for content management, EAI, and business process management will be integrated inside the ESA platform and will be made to work with the ecosystem of objects. Adapters to certain platform components for specialized functions outside the ESA platform will be created.

The objects at the application and generic services layer will contain sufficient data from the enterprise applications to support the first generation of PCAs. Stage 1 ESA platforms will have shareable objects in the application services layer for the most important data (such as customer, product, employee, and business partner data). Some of this data may be so standardized that it could reside at the generic services layer. A closer look at the objects will reveal that each cluster of objects is related to an underlying enterprise application. This is where the fact that the enterprise applications are carved into distinct silos will work in favor of the ESA platform architect.

Services for a Stage 1 ESA platform will come mostly from the platform components with some additional functionality to search for data and a few other functions from the enterprise applications. Some processes from the enterprise applications may be available to the PCAs, but PCAs will probably use business process management functionality to create the processes they need.

The development environment for the Stage 1 ESA platform will probably be similar to the development environment for the current generation of application servers. Programmers will be able to browse through the interfaces for the available objects, services, and processes and then use interactive tools to knit together the user interface and all of the underlying functionality. Developing in the Stage 1 ESA platform will be similar to coding in Java or C #.

The PCAs for the Stage 1 ESA platform will be able to access all of the data from all of the enterprise applications and will have a powerful collection of functionality inside the ESA platform with which applications can be created. The PCAs built on this Stage 1 platform may have to do things like present their own interfaces for certain types of configuration that may be more automated in future versions of the ESA platform. Figure 4-8 shows a typical Stage 1 ESA platform.

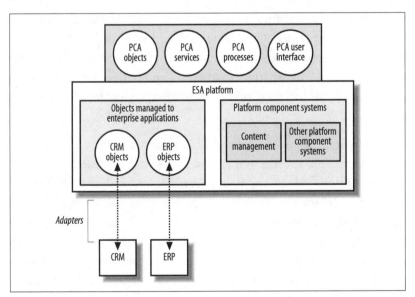

Figure 4-8. Stage 1 ESA platform

Stage 2: Standard Objects and Services

In Stage 2 of the development of the ESA platform, the data and supporting services are standardized in an object model. In Stage 1, the object model was built on top of the enterprise applications, with a set of objects that represented the data in each of the different enterprise application silos. If customer data was in two or three objects, it was the PCA's job to sort this out, perhaps by creating an object that assembled all of the customer data that it needed from the various objects from each of the silos.

In Stage 2, standard objects are created to make sense of all of the data in all of the silos. If customer data is spread out in three or four silos, then a customer object in the standard object model sorts this out and presents one customer object to the PCA developer. The services to manipulate these objects are also standardized, so at the application layer, services like "find expert" are abstractions of functionality of the enterprise applications, not merely gateways to their functionality. For example, in Stage 1, a PCA might use an application layer service to find an expert in a skills database of the HR application. In Stage 2, "find expert" would be a service that worked against the standardized object model which contained objects that had the expert information in them. The step forward may be a small one, but it has large practical significance for the ESA platform vendors.

The benefit of the standardized object model is that it eliminates dependencies in specific implementations of the underlying enterprise applications. At first, most ESA platform vendors will probably support only specific enterprise applications. For example, if an application suite vendor, which builds and sells enterprise applications, creates an ESA platform, it will likely come with adapters to its own enterprise applications. In Stage 1, the objects will represent the way that its applications present data. But in Stage 2, the standardized view of the data offers the possibility of supporting enterprise applications from different vendors or using legacy systems as the source for certain types of data. PCAs would be written to access the standard object models and the adapters to the CRM system from two different vendors would populate the standard object model. While it is possible to support two different versions of CRM systems without a standard object model, using a model makes solving the problem much easier.

If underlying enterprise applications have objects, services, or processes that can participate in the standard model, then they could be used at the application or generic services level without the need for an adapter. As time passes, more and more services should be available as part of the standard model from enterprise applications because web services will expose more and more functionality.

The standard model should also change how the development environment is presented to the PCA developer. The standardized objects and services should be the units available for use by the developer, who should benefit from the availability of these standardized components. Figure 4-9 shows a typical Stage 2 ESA platform.

Stage 3: Standard Processes

Stage 3 extends the standardization to processes. The reason this is a separate stage is two-fold. First, processes generally require objects and services to do the work of each step, so it is hard to define standard processes in the absence of standard objects and services. The second reason is that the enterprise applications will likely expose access to data, which will support objects, and access to services, which will provide ways to get at and manipulate the data, before they provide access to processes. This is likely to be true because of the relative ease of exposing data and services as opposed to processes. Processes are less atomic than objects and services and require a continuing conversation about what is happening as the process flows through each step, which complicates how they are presented for use. Some applications are exceptions and have the ability to allow some processes to be exposed in their current ver-

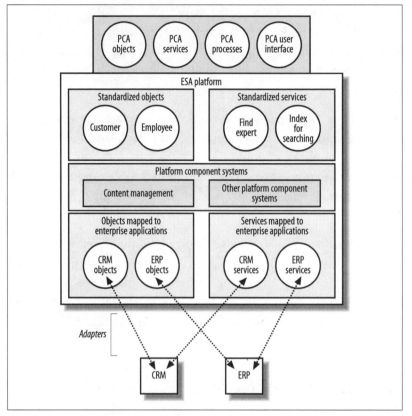

Figure 4-9. Stage 2 ESA platform

sions, but even these applications will have more processes available for use in future versions.

The processes that will be exposed from enterprise applications are likely to fit into the ESA platform at the application services layer. It is possible to conceive of a generic process like changing an address. Most of the time, however, people think of processes as higher level constructs that aggregate lots of functionality, which means they would not fit in the generic services layer.

The development environment will allow developers to knit together processes as components. With the exposure of standard processes, the developer should have a completely abstract view of the enterprise. Figure 4-10 shows a typical Stage 3 ESA platform.

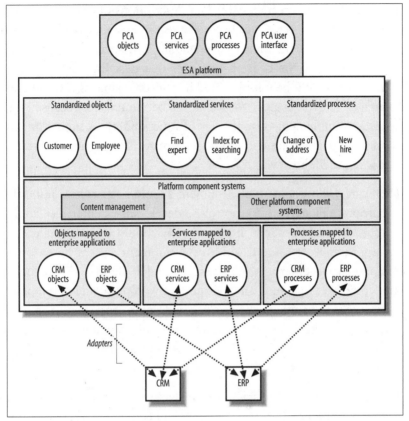

Figure 4-10. Stage 3 ESA platform

Stage 4: Model-Based Development

In Stage 4 of the evolution of the ESA platform, the standard objects, services, and processes are refined into a comprehensive enterprise application model. The leap forward in this stage is the way that the development environment uses this comprehensive model to allow developers to knit together an application in terms of a model.

In the previous stages, as the standard objects, services, and processes appeared, the development environment allowed them to be used. The advance in Stage 4 is the degree to which the model and modeling are the center of development activity, as opposed to coding. This evolution implies some large changes in how applications are described and created. Metadata descriptions of objects and the relationships between them become the primary means of controlling application structure. This metadata is then used to generate applications with the help of templates that describe how a certain relationship would be implemented in a specific language. Coding in languages like Java and C# will never be eliminated because it is needed to implement certain sorts of functionality. A model-driven development environment should minimize the amount of coding that is required, however.

The Stage 4 development environment should be attractive not only to corporate IT users, but also to the developers of enterprise applications. A whole new era of productivity should be inaugurated by the appearance of a model-driven development system based on a comprehensive enterprise application model. Besides the reduced time to market that such an environment would provide developers, the fact that the applications will live within the ecosystem of an ESA platform should make many more layers of components and services available for reuse. Figure 4-11 shows a typical Stage 4 ESA platform.

To the marketplace, the stages of the evolution of the ESA platform will be obscured by the fact that PCAs will come to market and offer applications that completely fulfill the goals of the ESA architecture before ESA platforms are completely developed. What the users of these PCAs won't realize is that first generation PCAs will implement functions in the PCA layer that will be included in later generations of the ESA platform.

Platform Components and the ESA Platform

Now that we have dissected the innards of the ESA platform, we will describe how the ecosystem will present itself to the PCA developer. In looking at this area we will cover in greater detail exactly how the platform components show up in the ESA platform.

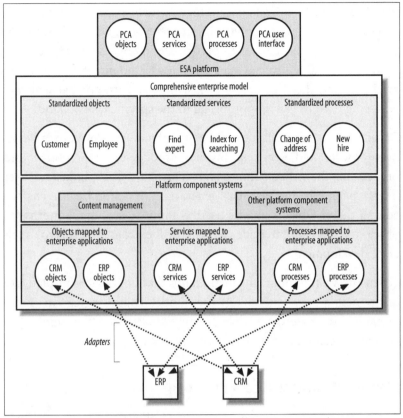

Figure 4-11. Stage 4 ESA Platform

The job of the ecosystem, remember, is to provide a way that objects, services, and processes can be created and interact with each other. As these become standardized, the developer is presented with the option of configuring them and choosing how they relate to each other. The way this is likely to happen is that a standard business object will be created. That business object will incorporate a multitude of standard services that the developer will turn on or off. The object will then be able to participate in the standard processes. Application development will consist of getting the basic structure of the application in place using the model to assemble the standard components and then adding the missing portions.

For example, when a developer creates a business object, he or she may see functions that allow the object to send and receive messages, to be indexed by a content management system or to invoke a search, to notify another object of an event or to wait for a notification, to invoke a process or to participate in one. When an object is created, it is likely to have an enormous amount of functionality available, and the developer's task will be to turn off that which is not being used or is not appropriate to the specific task at hand.

Different sorts of plumbing will be available to the developer to connect objects and services. Business processes will control and orchestrate the other components. Objects will be linked to user interface templates that help generate and customize screens for editing them. Objects will be used as input to services. Patterns for processes and user interface elements will exist at all levels to accelerate development.

Given the vast amount of functionality that is available in platform component systems, it is likely that modeling and aggregating functionality into patterns will be required to reduce the amount of detailed information each developer must keep track of.

Standards and the ESA Platform

Another implication of the standardized objects, services, and processes is that they become *de facto* standards for the enterprise applications and platform component systems. This standardization makes it easier for the ESA platform to support different versions of underlying enterprise applications or platform components because it provides an abstraction of the data or functionality provided.

But difficult questions also arise. How will extensions or customizations be handled? Who will define the standard? Will standard wars erupt around how the standard models are created?

Vendors are likely to view and react to the standards issue in completely different ways based on their interests, installed base, and relationship to the customer. In the next chapter we will discuss how PCAs will affect each type of vendor and also look at how PCAs are likely to make demands on the organizational, strategic, and operational dynamics of the enterprise.

5

Business Value and Strategic Impact of PCAs

When is a technology powerful enough to change behavior? If a technology changes behavior, what will be the impact? Who will benefit and who will lose? How will the drama unfold?

With something as simple as developing film, dramatic changes in behavior did not happen when the turnaround time was reduced from a week to a day. Two trips to the photo shop were still required. But when film could be developed in an hour, only one trip was required. Film could be dropped off at the beginning of a trip to a mall and picked up at the end. One trip was eliminated, and the model changed.

PCAs offer the potential to change current models of how technology is acquired and implemented in a way that may reshape the network of relationships that bind a company and its vendors. PCAs will be different from current vendor offerings. PCA products will offer both targeted solutions and broad solutions, solutions that implement a focused business process and those that cross many functional areas. Both sorts of solutions will unify information and services from heterogeneous systems and provide flexible business process management. PCAs will plug into a company's infrastructure like a USB device plugs into a laptop.

PCAs will change how vendors do business. PCAs will lead ISVs to start expanding the scope of their products by using additional services provided by the ESA platform. Systems integrators will be able to turn some of their consulting expertise into PCAs. The landscape for technology vendors who sell application servers and other integration systems will change because the ESA platform will make even more of the application stack a commodity. These trends attract a swarm of questions.

If the cost of innovation drops, does that result in more projects with the same budget or fewer projects with a smaller budget? If PCAs proliferate and provide plug-and-play business solutions with configurable processes, does this change who in a company buys technology? If business process management becomes the key task in unlocking the value of technology, where do the modelers come from? Do systems integrators provide PCAs or do they make their living creating custom composite applications? Do ISVs become PCA vendors or systems integrators?

To understand the impact of PCAs, their potential business value, and the way they will reshape enterprise technology, we will look in turn at how PCAs will impact suite vendors, systems integrators, and ISVs. But to begin we will look at PCAs and their effect on the corporation—on its technology and on organizational dynamics.

PCAs and the Enterprise

The unique characteristics of PCAs—that they leave existing systems in place, that they are smaller and more targeted applications, that they provide flexible processes—contribute to making their arrival less traumatic than that of previous generations of applications. If ERP systems were gorillas and enterprise applications like CRM were a troupe of chimpanzees, then PCAs would be a parade of ants.

PCAs will cost less than previous enterprise applications and will arrive with an ESA platform that may eventually become the foundation for all internal custom development. Both the focused products and the power of the ESA platform should work in favor of lowering the cost of innovation, as we argued in Chapter 3. PCAs will also change the perspective of those using the applications. Instead of presenting a small portion of a database on the screen of the user interface, the PCA will show a unified view of information collected from a variety of systems or the details of an end-to-end process that crosses traditional application and process boundaries. This expanded reach will broaden the audience for key information and change the consciousness of the organization, making everyone more aware of the whole than of their specific part.

PCAs and the ESA platform will also change how a firm thinks about its technology. Architects can imagine their systems not as large boxes of functionality—the gorillas and chimpanzees of the enterprise applications like ERP, CRM, or SCM—but instead as the hundreds of processes, services, and objects that those applications provide. Using models to manage the complexity and express the relationships, architects will be able to bring these elements together into useful components that can be reassembled, connected, and reconnected to reflect the strategy of the company. PCAs will be the first useful incarnation of these components, but eventually the IT infrastructure will be envisioned and manipulated in this manner with business process models as the glue that holds everything together.

When the swarm of PCA ants arrives, what will happen? How can companies benefit to the greatest extent possible? We will now answer these questions in the areas of strategy, operations, organizational dynamics, and vendor relationships.

Strategic Impact of PCAs

The problem with strategy is implementation, not vision. There is generally no shortage of ideas about new ways of running a business, attacking the market, and making customers happier. But there is a limited amount of time, money, and capacity to implement new strategies and frequently the biggest bottleneck is the cost of retooling the IT infrastructure.

PCAs will make strategy easier to implement in two ways. First, the initial wave of PCAs will take advantage of the unified view of enterprise information to create applications that improve strategic efficiency. Applications of this type, a few of which we discuss in Chapter 6, help automate the process of assembling and analyzing information and deciding what to do next. A PCA for product definition, for example, would collect ideas from a diverse collection of people inside a company, and then manage the prioritization, analysis, and evaluation of those ideas using information from all of the company's enterprise applications and the collaborative and business process tools provided by the ESA. The result is a faster process with less time spent assembling information or retracing steps to figure out how decisions were made.

Second, PCAs will provide options for changing strategy, creating solutions that form the basis for strategic process changes. Vendors will come to market with niche products aimed at urgent problems. In financial M&A transactions, for example, one of the largest problems is account consolidation. Any customer of a bank that has merged with another knows how frequently the consolidation process goes awry. A PCA focused on managing this process, with embedded best practices as well as the ability to flexibly adjust the process to battlefield circumstances, would have an immediate positive impact. It would be a strategic impact if the PCA allowed the company to change the way it does business.

The longer term impact on the range of strategic choices available to a company could come from letting internal development staff build custom composite applications on the ESA platform. Applications that have the highest strategic value will most likely be proprietary and are not likely to be provided by vendors. An application of this sort might be a proprietary forecasting and pricing model populated with data from all of the enterprise applications. Such an application provides a company with insight into how market forces will change demand for their products.

That fact that PCAs will cost less and have a faster ROI than enterprise applications should expand the practical choices available to an enterprise. The smaller price tag means more functionality for the dollar. The smaller scope means more rapid decisions to implement

PCAs. The power of the ESA platform used for custom development means systems integration budgets should go farther. As a result, innovation should be cheaper and automation should be extended to more processes. This is likely to increase the number of new projects that a company undertakes or reduce the cost of implementing a specific set of innovations.

The reduction in the cost of new functionality and the acceleration of implementation should also change the landscape of a company with respect to expanding partner relationships. A more supple infrastructure based on the ESA platform will allow partner services to plug into PCAs or custom composite applications. Supporting and changing relationships with supply chain partners will be much easier with the service-based approach of ESA.

If speed of decision making, faster implementation of strategy, lower cost for technology, and advanced ability to support partner relationships make a difference in a competitive environment, and almost all experience suggests that these factors are crucial, then PCAs should provide companies who employ them a competitive advantage. PCAs should illuminate a business like a three-dimensional graphic that can be rotated to show new perspectives and reveal information that was previously hidden. The velocity of change and an accelerated pace of innovation should eventually result in happier customers and put competitors on the defensive. Ideally, with a stronger position, a company becomes the aggregator, not the target.

Operational Impact of PCAs

PCAs should change the way that a company thinks about its operations. In the current world, the major components are large systems like ERP, CRM, and SCM linked together with difficult-to-change integration technology. Processes like order-to-cash are hard-wired into these applications, each of which may play a role in implementing the process. Changing these processes is expensive. Portal technology can bring together all of the information from the underlying applications into an interface that shows it all in one place. But to do anything, to take action, one must leave the portal and return to the original enterprise application.

In the world of PCAs, the architecture is no longer in such large boxes. It is homogenized by the ESA platform and appears as a unified model for processes, for services, and for objects. The existing enterprise applications do the work, but they are no longer visible. The PCAs and custom composite applications that business and technology architects have at their disposal contain a comprehensive view of the enterprise. This new vision of the infrastructure increases the importance of creative thinking about process. The question is no longer, "How can I tinker with the hard-wired processes that link my enterprise applications?" The lowering of the implementation barrier through ESA means the key question is now, "What process will produce the most business value?"

To address this question, PCA vendors will offer solutions that implement the best processes possible to handle tactical problems. Companies will then be free to implement processes that are not bound by the constraints of the current architecture.

The processes provided by vendors in PCAs or implemented by the company in custom composite applications will cross the boundaries of the organization, the enterprise applications, and the existing hard-wired processes. The operational impact of this architecture will be that more people have access to more information about the tasks they have to perform. Processes will be increasingly automated and collaborative functionality will support processes that are not automated.

In the oil industry, early versions of PCAs have allowed companies to automate the performance of asset teams, which are collections of experts from a wide variety of functional areas, assembled to focus on operating a large asset like an oil rig. One key aspect of these PCAs is a flexible process editor that allows an instance of a process of fixing or maintaining a piece of equipment to be started and then adjusted based on the needs of that particular job. Hand-offs between steps are handled through email and other alert mechanisms. Information flows where people need it, whether they are in the office or in the field on a rig in the middle of an ocean. The systems eliminate lag time between steps of the process and reduce the need for meetings because a comprehensive view of the state of the project is kept in the PCAs.

This example points to perhaps the largest tactical transformation that PCAs will bring about. Processes will no longer be described in a graphical tool like Visio and then implemented in code. The description and the implementation will become the same. Corporate tactics will mean shaping processes directly instead of shaping them by changing hard-wired connections between enterprise applications.

Another transformation will be a rise in using modeling to make sense of the infrastructure and the possibilities. In the ESA world, one has to keep track of a hundred-fold more processes, services, and objects that have been crafted out of the existing systems instead of a smaller number of possibilities in the existing systems themselves. Models of processes, services, and objects will become the equivalent of today's data dictionary, and modeling skills will be in high demand.

Organizational Impact of PCAs

Today's enterprise applications are hard-wired together to automate business processes. The rigidity of this architecture puts a strain on the organization. The tension results from the central pull of standardization and cost-cutting and the desire of each division to optimize the infrastructure for its own needs. The CEO wants one system, standardized to take advantage of economies of scale. The head of a business unit wants a custom system to improve efficiency. In general, the CIO sits in the middle of this dilemma and one or both sides is unhappy with the solutions provided.

The PCA paradigm offers both sides of this equation what they want. Placing an ESA platform on top of a suite of centralized enterprise applications derives additional value from an existing investment by bringing information and services to a larger audience. Using PCAs or custom composite applications to deliver functionality allows each division to get the flexibility it needs at a reasonable cost without disrupting current users. The complexity of this architecture is manageable because applications are primarily a recombination of existing functionality with a small amount of new functionality. If most of the functionality is provided by PCAs, then the vendor solves the maintenance problem.

A more flexible, customized infrastructure is just the beginning of the effects of the PCA and ESA architecture on a business. The largest impact of PCAs will probably come from how they change the way people think about their jobs and about the business at large. The availability of information in context will likely be the largest agent of change. Working through a scenario will make this clear.

In a product launch process, a variety of departments must coordinate their actions so that the product is ready to ship, the marketing materials are ready for the sales force, events are properly staffed, everyone is trained, and hundreds of other details relevant to launching a product are managed. With a master view of the schedule, it is possible for each team to know if it is ahead or behind and of the potential impact on the entire project. If something goes wrong, improved information means less room for finger-pointing and perhaps allows underutilized or overbooked resources to be identified early enough to forestall disaster. Each person in the process becomes aware of how he fits into the big picture. With more people taking a global view and communicating about it, better performance, better suggestions for improvement, and fewer mistakes are inevitable.

Let's look at the perspective of senior management on how a PCA might change their view of the situation. Instead of hearing summarized reports in status meetings about how the product launch process is proceeding, it should be possible, using a portal-like dashboard, to quickly look at all of the process steps that are behind schedule. The impact of the delays can be examined and more resources can be allocated or priorities can be shuffled to solve problems in advance.

The result of better, up-to-date information is the enhancement of central control and influence. Awareness of each part in a big process allows it to be managed. With better information and a more comprehensive viewpoint, management intervention will be more effective.

Another result of more information is that it replaces politics with performance. Politics is the management of perceptions and it fills vacuums in information. With better information about the state of each part of a larger process, there is less room for politics: tasks are either done or they are not.

From an investor perspective, this all sounds great. Better performance, lower cost of change, informed management, less political noise, and all of it should mean better execution and happier customers. But from a management perspective, there are some fascinating problems to be managed in the wake of the cross-functional processes enabled by PCAs.

For example, how is group performance rewarded? Is the group involved in a process the right unit for team incentives, even though it may cross boundaries on the organizational chart? Should the traditional divisional hierarchy for reporting give way to teams focused on various processes? Will a job focused on a specific function in a vertical hierarchy be replaced by a cross-functional role in a horizontal, process-focused team? If so, how is everyone managed and allocated for expense purposes? What is the career path? Will people rise in responsibility in the organizational chart or by playing a more senior role in a process?

Security and privacy come into play when allowing access to comprehensive views of the enterprise. How can information be filtered properly to allow someone playing different roles in different projects to access only appropriate information?

Entire job descriptions will likely change in the PCA paradigm. Will the CIO role change from that of a builder and integrator of infrastructure to that of a master modeler and broker of services, who chooses internal and external methods of provisioning the services needed by the enterprise? Will this be a more powerful or less powerful position? Given the focus on process and the plug-and-play nature of the infrastructure, is it possible that all technology will now be sold to the process owners, the heads of the business units?

Given all the forces in play, it is clear that PCAs have the potential to dramatically reshape a company's organizational dynamics as well as to expand the scope of available strategy and tactics. But bigger changes still may be in store for the vendors of technology. Before we examine the effect on vendors in detail in the next part of this chapter, we will take a quick look at how the changes in the vendor landscape will appear from the company's perspective.

Impact of PCAs on Vendor Relationships

Each of the four basic food groups of vendors—technology vendors, suite vendors, independent software vendors, and systems integrators—will use PCAs in a different manner to enhance their relationship with a company and increase their share of the budget. Each of these vendors will come calling on a company to sell PCAs and an ESA platform. Here's what they are likely to say.

What technology vendors will say about PCAs

Technology vendors are companies that sell core technology platforms like application servers, databases, content management systems, integration technology, and other products out of which applications are built. IBM, Microsoft, Oracle, BEA, Sun, and WebMethods are all technology vendors (although companies like IBM and Oracle may appear in more than one category). Technology vendors are likely to make the following case:

- PCAs and custom composite applications can indeed add tremendous value to an organization.

- PCAs will not really be products for quite some time, but companies can get the value of the approach now by building custom composite applications with our application servers and other technology that can bring together the entire infrastructure and expose it through web services as a homogeneous ESA layer.

- We are working with our systems integrator and ISV partners to create PCAs that will plug right into our development platform.

- Companies already have our technology, their developers know it, and so do the systems integration firms they work with.

The technology vendors will argue that this approach is the fastest path to the benefits of composite applications, and custom development will pay for itself quickly and will be of higher value than PCAs.

What suite vendors will say about PCAs

Suite vendors sell collections of enterprise applications that are aimed at automating the enterprise. Usually the center of the suite is the ERP application, which has a comprehensive model of the financial, organizational, and product structure of a company along with hundreds of optional modules for special purposes. Suite vendors also usually sell products for CRM, SCM, HR, and other applications that integrate with each other and with ERP. SAP, PeopleSoft, Baan, JD Edwards, and Oracle are all suite vendors. Suite vendors are likely to make the following case:

- PCAs based on an ESA platform can indeed add tremendous value to an organization.

- The PCAs we offer are made to work on top of a company's existing applications, which we also built, and will fit perfectly into a company's architecture.

- The ESA platform we offer can employ our applications or those from other vendors as the foundation for your PCAs. We are building adapters to unify the current heterogeneous environment.

- Our knowledge of all the enterprise applications has allowed us to make the most powerful ESA platform with the most comprehensive offering of processes, services, and objects.

- We are working with ISVs and systems integrators to create PCAs on our ESA platform. Companies, if they so choose, can develop custom composite applications using our ESA technology.

- PCAs are more affordable and cheaper to maintain than custom development.

The suite vendors will argue that their deep understanding of how to create and support enterprise applications and all of the functionality they have available to them from their current enterprise applications give them the ability to actually make PCAs work as products, which will avoid the expense of custom development. For projects that really must be customized, a company can write custom code on the ESA platform.

What ISVs will say about PCAs

Independent software vendors sell enterprise applications or technology platforms with specialized functions. ISVs usually succeed by being the first company to create a certain type of application or by having advanced functionality suited to the needs of a particular industry. Companies like Seibel, Documentum, Best, and Verity are all ISVs. ISVs will likely make the following case:

- We think that PCAs make it possible to unlock the value of our application by lowering the cost of integrating with other systems.

- We are going to support the major ESA platforms so companies can use our functionality as a component to create custom composite applications.

- We are writing our own PCAs and also working with other vendors to include our functionality as a component in their PCAs.

ISVs have had to beat suite vendors, technology companies, and systems integrators by convincing companies that their product was better than competing products or custom development. The ISVs will continue to argue the same thing using PCAs as a different way to package and deliver their products.

What systems integrators will say about PCAs

Systems integrators sell services to create custom solutions and integrate software products into a working solution. Systems integrators succeed by developing skills and expertise in areas where companies have needs and by executing on projects to create business value. Companies like Accenture, Bearing Point, and CAP/Gemini are all systems integrators and will likely make the following case for PCAs:

- PCAs and custom composite applications are indeed the next wave for IT and have tremendous potential to create business value.

- The ability to use the ESA platform will increase the productivity of our traditional engagements to create custom applications.

- PCAs will require a fair bit of customization and the most value will come from custom composite applications, based on an ESA platform, that integrate tightly with a company's legacy applications and include some proprietary functionality that the systems integrators bring to the table.

- For certain products that may have a broad appeal, a systems integrator may work with a suite vendor or an ISV to productize a PCA.

Systems integrators will use PCAs as a launching pad to sell more services, as they have with every new wave of technology.

Now that we have looked at what the vendors will probably say about PCAs to their customers, the next section will examine in greater detail how PCAs and ESA will affect each type of vendor and the strengths and weaknesses each brings to the arena.

How PCAs Will Transform IT Vendors

If vendors have been selling and caring for gorillas and chimpanzees, what will happen when they try to sell ants? Each type of vendor holds a substantial claim on software and skills that are vital to the successful implementation of enterprise IT, but each group also has significant challenges to overcome to succeed in the PCA/ESA world.

In this section we will examine the strengths and weaknesses of each class of vendor with regard to exploiting the PCA/ESA paradigm. We will look at the assets and skills present and lacking in each class and speculate about competitive strategies. We hope to better understand how vendors will increase revenue, how they will avoid becoming a commodity, and how they plan to eat each other's lunch.

The victors among the technology vendors, suite vendors, ISVs, and systems integrators will be the companies who use PCA/ESA to create more value for their customers and use their products and service offerings to lead the transformation that the PCA/ESA paradigm will bring about.

Governing Forces for Vendors

If technology vendors made out like bandits in the Internet boom, they are paying a heavy price now. Technology spending has decreased across the board, companies cast a jaundiced eye on vendor claims for products, and a terror-stricken look comes over executives who have to face a custom development project. So much technology was purchased unnecessarily or failed to meet expectations that for the past two or three years most companies funded only projects that could produce a positive ROI within six to nine months. Technology innovation has become a zero-sum game.

This drought is part of the ebb and flow of capitalism and is not altogether unwelcome to the long-term participants in the industry who had to face irrational, incompetent competitors who frequently left clients hating all vendors after colossal and expensive failures. The reward for the survivors is less competition in the next wave of transformation of IT, which this book argues will be led by the notion of the PCA.

PCAs, from the vendor perspective, satisfy a basic need for new products and services to sell. Each wave of growth—from mainframe through client/server and the Internet—satisfied needs, provided infrastructure that became a commodity, and then created new opportunities. As the IT industry matures, the search for opportunity becomes more difficult. Companies are full to the brim with technology solutions that they are struggling to understand, implement, and optimize. Revenue is further squeezed by international competition fueled by low-cost labor, new entrants to each class of vendor, and the increasing viability of open source technology, which is free.

The trend toward commoditization and the pace at which it occurs is also a large problem for vendors. Everyone buys commodity infrastructure, but very little profit is made in selling it.

In the sales pitches at the end of the last section, we saw the face each vendor will probably present to the customer. Now we will attempt to explain the motivation for each viewpoint and its relative ability to deliver.

As a way to compare each class of vendor, we will use the ESA architecture as a yardstick. Each vendor will be evaluated on its ability to deliver the functionality at the application services and generic services layers, as well as how well they will be able to create and support PCAs. We will use the categorization of these layers found in Figure 5-1.

We will start with technology vendors, then move through suite vendors and ISVs, and end with systems integrators.

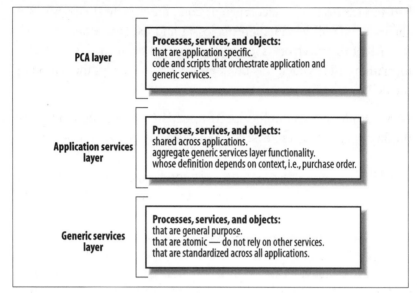

Figure 5-1. Definition of ESA layers

Technology Vendors

Technology vendors know how to build software that others use to build applications. Operating systems, application servers, databases, and core infrastructure like messaging systems are the most complicated programs currently on the market—and the most difficult to support.

The companies in this class—IBM, Sun, Microsoft, Oracle, BEA, WebMethods, and others—are all great at building and supporting complex products. Technology vendors generally have huge staffs devoted to quality assurance, release engineering, documentation, and support, and many of them have decades of experience and finely tuned processes to manage the complexity of the task of building hard-core technology infrastructure. Most applications are far easier to build than infrastructure, so why shouldn't the technology vendors be the toughest competitors on the block when it comes to creating PCAs and an ESA platform?

At the generic services layer of the ESA platform, technology vendors will be fierce competitors. The generic layer's job is to communicate with the existing enterprise applications and legacy applications to allow those programs to be used as services by PCAs. The generic layer is just like technology infrastructure in that it is completely general and not at all oriented toward a specific business. Lots of the plumbing that will be required in the generic layer has existed for years in current versions of application servers and other products or is the focus of large development efforts. Technology vendors have deep knowledge of how to connect applications via APIs, XML messages, or web services.

Technology vendors already offer Integrated Development Environments (IDEs) and

Software Development Kits (SDKs) that enable programmers to build applications. Each of the large technology vendors has a large developer community that has spent years learning their technologies. This community is a large asset and technology vendors invest millions each year to communicate with developers and keep them abreast of how the vendor plans to improve its software.

Knowledge of the enterprise applications is the large barrier at the generic layer for technology vendors. In ESA, much of the work is done by existing applications already in place at a company. The ideal ESA platform allows any CRM system to be the basis for the CRM-oriented services offered by that application. The technology vendors do not generally offer their own enterprise applications and do not have a deep understanding of the nature of each of the enterprise applications in-house. This deep understanding is crucial to the task of breaking up the underlying enterprise applications into a set of services. Technology vendors do understand platform systems like content management and would know how to break those up into component services.

At the application services layer, technology vendors enter new territory. This layer is application-specific, configurable, and messy. Application layer processes, services, and objects are constructed out of many different generic services or from services provided by enterprise applications and channeled through the generic layer. The application layer is not standardized like the generic layer. Shared objects change based on the business context. A purchase order for one company may be substantially different from a purchase order at another company. The configurability and amorphous nature of shared objects is more akin to the kind of configuration built into applications than anything in the generic layer.

At the PCA layer, the technology vendors have a huge asset—the skills and staff to support and maintain installed software products. This is a large leg up on a potential barrier to entry for the PCA market. The new territory for technology vendors would be the creation of PCAs for specific vertical markets. Successful products are founded on an in-depth knowledge of the customer's business. Gaining such knowledge is hard work. Most large technology vendors do have systems integration divisions that spend significant time with customers. It is likely that technology vendors would sell PCAs defined by their systems integration divisions. It is also possible that systems integrators could bring product ideas to technology vendors who would then market and support the PCAs through a partner arrangement.

The path of least resistance for a technology vendor would be to extend its application server or related technology into an ESA platform and then go into partnerships with suite vendors for the underlying applications and with ISVs, systems integrators, and suite vendors for ideas for specific PCAs. The problem with this approach is that while it plays to the strengths of technology vendors, it keeps them focused on the layer that will be commoditized soonest. If technology vendors can find a way to design and build PCAs, they can start eating into the market for suite vendors, systems integrators, and ISVs, which could encourage these groups to base their offerings on the technology vendor's platform.

Competitive issues like this complicate the relationships between technology vendors, suite vendors, ISVs, and systems integrators. We will analyze these complications at the end of each of the following sections.

Suite Vendors

Suite vendors know how to build and support enterprise applications. Suite vendors make a living by understanding the needs of businesses in various vertical markets and then creating general-purpose products that appeal to the largest possible number of customers.

The companies in this class—SAP, Oracle, PeopleSoft, Baan, and JD Edwards—all have a significant installed base of their main offering: the ERP systems that act as a central hub for the flow of information in large companies. Radiating out from ERP is a host of enterprise applications used by CRM, SCM, HR and others. Processes are automated by having a bit of the process done in SCM, another bit done in ERP, and another bit done in HR, and then having all of the applications hard-wired together. This methodology represents the state of the art for process automation at most large companies. Hard-wired processes are expensive to change, and that is one of the main motivations for creating PCAs. With control of enterprise applications and deep knowledge of their client base, suite vendors should be formidable competitors in the PCA market. The key question is which ESA platform they will use.

The generic services layer shows the strengths of the suite vendors and also presents their biggest challenge. Suite vendors are good at building plumbing to make one application talk to another, and their complete understanding of enterprise applications should give them an advantage in breaking the underlying applications into services. Suite vendors are also excellent at defining, building, releasing, supporting, and maintaining products. They have fully developed product support organizations and processes.

The challenge for suite vendors will be opening up the internal structure of their technology. For most enterprise applications, suite vendors offer an SDK that allows developers to interact with their application through an API. This SDK controls the behavior of a portion of an application and generally stops far short of the sort of toolkit that technology vendors create to allow developers to build applications. To properly create an ESA platform, the suite vendors will have to create a far more fine-grained API than they have in the past.

Suite vendors will also have to present their expanded API within an interactive development environment. Suite vendors do have development environments today but they are not as comprehensive as those of the technology vendors, nor do they enjoy wide adoption among developers. While it is unlikely that a development environment will become a competitive advantage, to gain acceptance, a suite vendor's offering will have to offer, at minimum, a least common denominator of functionality present in current IDEs. If suite vendors can achieve this level of functionality, developers will have powerful economic reasons to build software for the suite vendor's ESA platform: doing so opens the door to the suite vendor's large installed base.

Suite vendors have significant developer communities for their technology both within companies and within systems integrators, but their reach is smaller with regard to developers who create new products. Few developers of new products start with an existing application from a suite vendor as the foundation. Most ISVs have had to interact with suite vendor APIs to build plumbing that enables their applications to communicate with those of the suite vendor, but this is substantially different from using that API to create an application. Another worry is competition; developers who use the ESA platform to create PCAs may face competition from the suite vendor.

The suite vendors have tremendous experience in the application services layer, where most of the functionality for their products resides. Getting the objects in this layer right is the key to building applications that appeal to the widest possible audience. Defining a product that incorporates the most knowledge about a customer's business problem is a difficult task. Allowing application layer objects to be configured and extended while still allowing them to be maintained as a product is another difficult task. Suite vendors have significant experience in addressing these vexing issues. Suite vendors also know how best to expose the functionality of enterprise applications as services at the application layer.

The challenge for suite vendors at the application layer will be in supporting each other's applications. The ideal ESA platform will allow any CRM system to provide services for the generic and application layers. The suite vendor knows its own CRM in detail and knows the general duties a CRM system must perform, but how easy will it be for one vendor to support another suite vendor's CRM through the ESA platform? This problem is worse for technology vendors, who do not control any of the enterprise applications yet must write an ESA platform that supports them all.

The PCA layer is another area of strength for suite vendors, whose business is founded on understanding business needs and creating products to meet them. It can be argued that systems integrators may have a better understanding of the kind of niches that will become PCAs, but it is a short trip from the sort of products that suite vendors offer now to the more narrow and focused products that will be PCAs. Suite vendors also have large product support organizations that are tuned to supporting the needs of applications.

Suite vendors will succeed by building compelling PCAs and encouraging systems integrators and ISVs to use their ESA platform to build them as well. Perhaps the worst case for a suite vendor is if the ESA platform of another suite vendor becomes dominant. A far less troublesome scenario is one in which the ESA platform of a technology vendor wins in the marketplace. In such a case, the suite vendor can switch to that platform and use it to create PCAs.

The trickiest relationship for suite vendors to manage may be with the ISVs who want to expose their functionality within the suite vendor's ESA platform. We will examine ISVs next.

ISVs

Independent software vendors tend to be smaller companies that succeed by building the absolute best solution for a specific problem. Some ISVs create applications like CRM or SCM. Others create platforms for functions like content management or search services. To survive in such a world, ISVs are generally nimble companies with a laser-beam focus.

Companies in this class include Documentum, Verity, Best, Seibel, and a host of smaller firms. They are finely tuned organizations that are very efficient at keeping their products up to date. ISVs usually have a consulting arm and a lean product support organization. Generally, once they have succeeded, the larger suite and technology vendors come after them quickly. Technology vendors offer competition with platform systems and suite vendors create competing enterprise applications.

While the revenues of the suite vendors companies are all within a factor of 10 of each other, revenues of the larger and smaller ISVs may differ by a factor of 100 or more. In this discussion, we will distinguish between two types of ISVs: smaller firms seeking to establish themselves and larger firms seeking to defend their position. The key question for ISVs as a class is how they will maintain the value of their unique functionality under the PCA paradigm.

At the generic services layer, ISVs are unlikely to be strong players. The largest ISVs have had to surround their core functionality with a significant amount of plumbing to allow for configurable processes, to connect with other applications, to solve problems with application servers, and to provide functionality that may be lacking in an enterprise environment. But even the largest ISVs probably will not want to fight the uphill battle of offering their own ESA platform. Smaller ISVs do not have the ability to play at the ESA platform level.

But ISVs have a tricky decision to make at the generic services layer. Just as ISVs have chosen to support specific application servers or to integrate with certain enterprise applications, ISVs will also make a bet on the ESA platform that they suspect will succeed and provide access to a new group of customers. The dynamics of this decision are different for smaller versus larger ISVs. Smaller ISVs will pretty much be happy to work with anyone who can get them more customers. Software from smaller ISVs is less complex, so the cost of supporting a new ESA platform is manageable, and they do not have to throw away any of their existing functionality.

But for larger ISVs, supporting an ESA platform will be a much larger cost, akin to deciding which databases or which application servers to support. Supporting an ESA platform also means abandoning much of the platform-like functionality that has been developed. The most frightening part of this for a large ISV is that supporting an ESA platform generally means a dependent relationship with a potential competitor. Both technology vendors and suite vendors, the likely creators of the ESA platform, represent the most consistent and dangerous competition to the large ISVs. It is possible that in exchange for supporting a particular ESA platform, the large ISVs will become the official provider of their particular functionality for the ESA platform, but this is the beginning of commoditization, in which their software is bundled and not sold on its own. Such a move could cannibalize their revenues.

Another issue for ISVs large and small is competitive pressure between ESA platform vendors. It may be too expensive for an ISV to support all ESA platforms. If an ISV picks an ESA platform vendor that is different from the technology vendors that it has traditionally worked with, will that reduce the cooperation from the losing platform vendor?

The application services layer is a fascinating battleground for the ISVs that reveals one of the largest issues that will have to be addressed in an ESA platform: the need for standards in the application services layer. For most ISVs, the value of their products for application developers is a large amount of tightly integrated processes, services, and objects that allow complex applications with specialized functionality to be rapidly assembled. The amount of ISV functionality in the application layer changes dramatically from company to company. ISVs that create applications have more in the application services layer. ISVs that offer platform systems like content management or search have more in the generic services layer.

The more an ISV's functionality is in the generic services layer, the easier it will be to expose the functionality for use in a variety of applications.

The difficult question is how ISVs' shared processes, services, and objects become integrated in the application services layer. The same question must be answered for enterprise applications and represents one of the most difficult design choices for the architects of the ESA platform. If the ESA platform chooses to use the shared objects of the ISV as a standard, the ISV benefits but the rest of the ESA platform must support that standard. For example, an ISV selling a specialized CRM system for the real estate industry may have a set of objects at the application layer that represent the different parties in a real estate transaction. Could these objects become a standard for this sort of application? If a standard set of shared objects is defined, how much of the ISV's shared functionality is determined to be part of the standard and how much will be deemed extensions? How will extensions be handled? Handling extensions is a complex issue because using them means that PCA code will be more complex and implementation-dependent. If the application services layer functionality is abandoned, what is left of the ISV's functionality? If standards are set for the ISV's functionality, does this increase the threat of commoditization? All of these questions can be answered definitively only for a specific ISV with respect to a specific ESA platform vendor.

At the PCA layer, ISVs should be strong competitors if they can find a way to integrate into the ESA platform. The competitive advantage of ISVs is a complete understanding of a specific application area or certain type of functionality. To survive, ISVs have to be able to translate this expertise into compelling solutions to business problems. The PCA layer is where ISVs should shine.

One attractive choice for smaller ISVs may be to keep most of their functionality in the PCA layer and to use ESA functionality to provide as little as possible. In this scenario, the ISV looks on the ESA platform as an enhanced application server. For large ISVs, this choice is less viable because their functionality is spread across the PCA, application, and generic layers.

But one large opportunity at the PCA layer awaits both large and small ISVs. ISVs excel at envisioning and building applications and own a certain sort of functionality that gives them an advantage. Could this comparative advantage make ISVs the application architects in the PCA paradigm? Instead of having to be limited by the amount of functionality they have been able to build, ISVs working with an ESA platform can now take advantage of a much larger set of functionality to create applications. By leveraging the support mechanisms of the ESA vendor, it is easy to imagine an ISV defining and building 10 times more applications on the ESA platform than they could afford to build with existing architectures.

Given the range of ISVs, it is hard to draw firm conclusions. But the forces we have examined suggest that ISVs strongest in the generic services layer are the ones most likely to ally themselves with ESA platform vendors and that ISVs who create applications will probably keep their functionality at the PCA layer.

Systems Integrators

Systems integrators sell services to help build and implement enterprise applications. Systems integrators provide the glue that fills in the gaps in an implementation, and in playing this role they get a microscopic view of the way their clients do business. Almost no software is sold without consulting services. The larger the project, the more likely it is that a significant amount of services are part of the sale. For the largest projects, consulting services are frequently a multiple of the software licensing costs.

Companies in this class include Accenture, Cap Gemini, IBM, Cambridge Technology Partners, and many other firms of all sizes. The services they offer include programming, architecture, design, business process consulting, documentation, project management, modeling, and anything else that is in demand. Systems integrators are great at picking up on new trends and selling their clients on ideas for new technology or on retooling existing technology. Systems integrators also tend to focus on specific vertical markets and become experts in the business processes and technology infrastructure of those markets. To systems integrators, PCAs represent a new opportunity for projects and perhaps a way to productize some of their expertise.

One fact changes the nature of the opportunity that PCAs present to systems integrators compared with all of the other classes of vendors mentioned so far. Systems integrators do not have product support organizations for packaging, releasing, and maintaining packaged software products. Systems integrators are tuned for selling consulting services; to become a vendor of products would require building a new sort of organization and adopting a different business model. Most systems integrators recognize how large a change this would be and are wary of the idea of becoming companies who sell products. As we will see later in this discussion, that does not mean that systems integrators will not be a factor in defining and implementing PCAs. It just means there is a high bar for them to take on the role as primary vendor and maintainer of a packaged application.

At the generic services layer, systems integrators will select which ESA platform to support as a practice area for services. A key part of systems integrators' value proposition is the expertise they bring to a problem. They gain this expertise by investing in training, performing pilot projects, and working with clients. The systems integrators will choose to support the ESA platform that they believe will bring in the most work.

The application services layer offers more opportunity for systems integrators. In many cases, systems integrators bring software to an engagement in addition to their expertise. This software most often implements processes, services, or objects that perform a difficult task or one that is specific to a vertical market. If a systems integrator has enough of these components, they may become the foundation for one or more PCA offerings.

Systems integrators should have the most influence at the PCA layer. Their knowledge of the client is even deeper in most cases than that of ISVs and should provide the foundation for defining PCAs that solve problems for client companies. The question is not whether systems integrators can define PCAs, but rather whether it is in their interest to do so. The main currency of systems integrators is their ideas and their ability to execute on those ideas. The question that systems integrators will take a hard look at when deciding to turn their expertise into a PCA is whether this means they are selling their ideas too cheaply. The deep vertical knowledge of systems integrators enables them to define PCAs for precise niches, and the smaller the niche, the smaller the total market. To justify creating a PCA, systems integrators will have to believe that they will make more money from license fees and related consulting services than they would from delivering their expertise as a custom composite application. They will also have to decide how the PCA will be supported, and if that means working with a technology vendor or suite vendor to provide support, how much money will be left for the systems integrator? Selling ideas for PCAs to someone else to develop and implement is another way to make money from PCAs, but is it the way to make the most money?

While some PCAs will no doubt come from systems integrators, it is more likely that PCAs and custom composite applications will primarily be a source of consulting services revenue. If a company adopts an ESA platform but has a large amount of legacy applications, then consulting services will be required to expose those applications in the ESA platform. If the enterprise applications in a vertical market are not yet exposed as services, consulting services may be required to bring them into the architecture.

What if ESA platforms are so easy to use that companies can do their own services work? This is a potential threat for ISVs, and one that has reared its head in the past concerning other technologies. Perhaps this may be true in some contexts, but the world of PCAs also seems replete with complexity, and managing complexity is the bread and butter of systems integrators.

Implications of PCAs for Enterprise Architecture

Enterprise architecture is the CIO's high-level plan for meeting a company's needs for information technology. The enterprise architecture is a set of principles, a set of tradeoffs, and a set of specific choices that describe how a CIO will respond to requests from business divisions. PCAs will change the face of enterprise architecture.

Right now, when CIOs look at their current infrastructure, they generally see an ERP application in the center of a complicated nest of legacy and vendor-supplied enterprise applications. Some of these applications are linked through point-to-point connections and others are brought into a unified interface through portal technology. Enterprise-wide processes such as order-to-cash are automated through hard-wired point-to-point connections that move information between applications like ERP, CRM, SCM and others. CIOs can think about business processes all day long, but their ability to do something about them is limited by the cost of changing the current architecture. CIOs can imagine a flexible supply chain in which suppliers are more tightly integrated, but creating such a supply chain is a lengthy and costly project.

PCAs and the ESA platform break the shackles inherent in the current architecture. The ESA platform provides a unified, homogenous view of a company's infrastructure. PCAs provide packaged functionality on top of this platform. Now, when CIOs look at their infrastructure they see the existing applications supplemented by a unified view of the infrastructure. The ESA platform provides an organized set of processes, services, and objects, in the application and generic services layers, constructed out of existing applications that can be combined and recombined at a lower cost than the traditional point-to-point integration. Change is built in to this architecture, to borrow a phrase from advocates of business process management systems.

But what does this mean for the role of CIOs and the contours of enterprise architecture? The largest impact may be on increasing the business focus of the CIO. The ESA platform productizes the sort of integration tasks that currently take a huge amount of a CIO's time. By lowering the implementation barriers to new processes, the ESA platform increases the time that a CIO spends on helping other executives figure out what the right sort of processes are. Once the right process has been designed, the CIO must implement that process, not with integration, but with PCAs and processes, services, and objects from the ESA platform that are built to be reconnected and configured. Bringing external partners into a company's processes becomes a matter of presenting their services in the ESA platform.

The ESA platform trades one kind of complexity for another. Replacing the complexity of the functionality of enterprise applications, integration technology, and portals is the complexity of the numerous processes, services, and objects of the application and generic services layers of the ESA platform. The CIO will become the keeper of the master model of the ESA platform and all of the applications that are exposed through it. It is one thing to expose applications as services; it is another thing to comprehend the semantics of those services—what they mean, how they change the underlying state of databases, and how they interact. Deploying PCAs and creating cross-functional processes will result in a multiplication of the complexity being managed, which will no longer be contained within application silos. Modeling skills will help manage this complexity, and the CIOs of the near future will likely be master modelers.

More configurable processes in PCAs and custom composite applications will finally give businesspeople the ability to control more of their own infrastructure. Many technologies have promised to wrest control of the IT infrastructure from programmers and give it to business analysts. PCAs and custom composite applications will gradually give more flexibility to businesspeople, not in one big bang, but gradually, one application at time. Will this mean the death of programming? Not at all, in our view. The gorillas, the chimps, and the ants—the ERP systems, existing applications, ESA platforms, PCAs, and custom composite applications—will all need care and feeding, but they will have a different ringleader.

6

Business Scenarios

PCAs spring to life in the computers of developers, but they succeed or fail in the business world. The graveyard of Silicon Valley is crowded with technology that was "better" than that which became commercially successful. Many a tear has been shed lamenting the fact that functionality is unimportant unless it creates business value.

In this chapter, we perform a thought experiment about various real-world scenarios in which PCAs will likely appear. Our goal is to discover the ideal conditions in which PCAs will thrive. Our method is to start with what we have learned about the forces in the IT marketplace and the capabilities of PCAs. We then conjure up some real-world situations that should help explain where PCAs will shine and where they will fall flat. For each of several business contexts, we will also point out key implementation details that will be crucial to success for PCAs offered in those contexts.

What PCAs Do

To prepare for our tour of various real-world situations in which PCAs are likely to be effective, we will review the structure and typical capabilities of PCAs.

PCAs are the application layer of Enterprise Services Architecture. PCAs are constructed from processes, services, objects, and persistence provided by the ESA platform. The ESA platform creates a consistent, unified, homogenous view of functionality offered by underlying enterprise applications such as CRM and platform component systems such as application servers and content management software.

PCAs are designed to solve specific business problems. PCAs do this primarily by combining core capabilities of enterprise systems. They also add new functionality that is specific to a particular problem. But in order for PCAs to successfully leverage existing systems, functionality and information from those systems should do most of the work. PCAs make use of existing applications and have the ability to tightly integrate them.

Here is a summary of PCA capabilities:

PCAs aggregate information from existing layers.
All of the information needed from underlying enterprise applications like CRM, SCM, and ERP is unified in a PCA.

PCAs create new relationships among pieces of information from existing layers.
The unified information from different underlying systems can be connected. Skills information from the HR system can be related to project information from project management software.

PCAs aggregate and generalize services from existing layers.
All of the operations that can be performed on a customer, for example, appear as a unified set of services, even though they may be performed by a variety of different systems.

All of the operations from platform component systems, like creating a process or searching for documents, are generalized for the PCA and apply to all objects in the ESA. The scope of a search, for example, would include objects from all enterprise applications.

PCAs aggregate processes from existing layers.
Processes that may be granular in underlying enterprise applications are unified in a PCA. A PCA might have an "available to promise" process that would look into the CRM, ERP, SCM,

and PLM systems to make sure a product was available and then record the fact that it was promised to a customer.

PCAs have discrete functionality.

PCAs are not monolithic, with a huge collection of related functionality, but rather are targeted to a specific function or process.

What PCAs Must Do Well

The functionality mentioned above provides a powerful foundation for PCAs to create business value, but it is not sufficient for success. PCAs must have three qualities in order to thrive as products in the enterprise environment. These qualities may represent the minimum cost of entry for new applications of any sort, but they are so important that they should not be taken for granted.

PCAs must be configurable.

The ESA platform sits on top of the existing infrastructure and provides a homogenous, unified view. But no matter now well that job is done, any particular installation may need an extra field of information or want to invoke a process or service in a legacy system. The more configuration rather than custom coding is used to meet these requirements, the better for the PCA.

PCAs must allow flexible process management.

In orchestrating processes from underlying applications and creating new ones to automate processes that cross the boundaries of application silos, it will be hard to predict in advance the sort of processes that will be required. When breaking new ground with automation, it is also wise to describe a process in a simple way and then let collaborative tools fill in the gaps until the process is better understood. If a PCA does not have the ability to allow configuration of processes and requires custom coding or difficult configuration, then the PCA will have limited value.

PCAs must integrate well with end-user applications.

The understanding of a process frequently begins in a spreadsheet that keeps track of the details. Collaboration begins with email. When a system arrives to automate a task, frequently the end-user applications have become entrenched as an efficient method of initial data capture. Introduction of a new interface must deliver a significant payoff. The more that PCAs elegantly move information in and out of end-user tools, the happier users will be and the greater the chances of success.

Ideal Conditions for PCAs

PCAs can be operational or strategic. They can be applied to brief or long running processes and workflows. As a technique, their power is very broad. Where will the functionality of PCAs described so far be optimally applied?

Having understood the capabilities of PCAs, we can now narrow down the universe of solutions that will be most appropriate to be implemented as PCAs. The ideal situation for a PCA is one in which:

- The current environment contains enterprise applications including any or all of the following: CRM, HR, ERP, and SCM.
- The current environment is heterogeneous across operating systems and enterprise application vendors.
- The current process is delayed by a large amount of manual processing to roll up information and perform other unification functions.
- The current collaborative processes need but do not have data from the enterprise applications.
- The industry is highly competitive and technology is a strategic factor.
- The solution can be provided for the most part with the processes, services, and objects from the ESA platform. Ideally, less than 25 percent of the functionality is provided by new processes, services, and objects.
- The solution requires a unified view of information that is currently stored in a variety of enterprise applications.

- The solution requires creation and maintenance of new relationships between data stored in existing enterprise applications.

- The solution requires a new cross-functional process in which actions are taken to invoke services of multiple enterprise applications and change the underlying state of the information.

- The solution requires unstructured information to be created, captured, and collaborated on.

- The solution requires integration of information currently stored in end-user applications.

With a knowledge of PCAs and the situations in which they thrive, we will examine some specific business scenarios to see whether they will benefit from a PCA and what that PCA should do to help.

A Tour of PCA Business Scenarios

We have chosen to examine four business scenarios, all of which have unique characteristics that should test the ability of PCAs to handle the complexities of the real world. The scenarios we will examine are:

Project Management
> The challenge of aligning a large portfolio of projects with corporate strategy, optimizing the allocation of resources and people to projects, and tracking performance.

Product Definition
> The challenge of assembling a large amount of ideas and requirements from diverse sources, evaluating them, selecting a short list to develop further, and carrying the ideas that show the most promise through the rest of the product development process.

Mergers and Acquisitions
> The challenge of helping mergers and acquisitions achieve the overall merger objective, despite the involvement of different teams in different project phases.

Employee Productivity
    The challenge of enabling employees to increase productivity and job satisfaction by automating repetitive and annoying tasks.

Project Management

Project management is frequently thought about in the context of a single process. In this examination, we consider analyzing a portfolio of projects.

Scenario: Myriad projects managed individually

In a large organization with thousands of people and hundreds of projects, a comprehensive view of all projects in all divisions is hard to come by. The most complete view generally arrives once a year as part of the budget review process, during which existing projects are reviewed, approval is granted for new and continuing projects, and some projects are cancelled.

The project managers who drive these projects forward keep track of them with end-user project management tools, spreadsheets, and other collaborative tools such as email and shared folders. Status reports for projects are rolled up at various points in the organizational hierarchy into higher-level summaries on a monthly or quarterly basis. Metrics on the rollup may include factors such as percent complete, days ahead/behind schedule, estimated delivery date, budget variance, and other such metrics.

By the time the information gets to the desk of the CEO or the committee responsible for aligning projects with current corporate strategy, the information may be months old. If more detail is required on a particular project, the project manager must be contacted to provide the additional detail. It is generally impossible to examine in real time or even with a two-week lag the entire portfolio from different dimensions such as business unit, project type, resources used, expected revenue impact, expertise allocated, and expected completion dates.

One problem with the current process is a general lack of visibility into the current state of projects and the difficulty of assembling all the relevant information from the different systems in which it resides. Employee skills information is generally part of the HR system. Customer information may reside partially in the CRM system, the financial applications, and in the SCM system. Specifications may reside in the PLM system. Project status is in end-user project management files that may not be integrated with a central repository of project information. Different divisions may use different end-user project management tools.

Another difficulty of project management at most companies is the endless parade of status meetings on projects at various levels of the organization. Decision-making is slow in such a world. When strategic priorities change, it generally takes a long time to figure out which projects still make sense. When projects start to fail, it generally takes a long time for that failure to result in a decision to cancel.

The Standish Group outlined the magnitude of project management failure in a 2001 report based on research of IT projects. A third of the projects they studied failed outright and were cancelled, creating an estimated $80 billion in losses. Just under half of the projects fell significantly short of their goals, ran over budget, took longer than expected, and ran up another $55 billion in unplanned costs. Less than a quarter of projects met their goals, on time and on budget. The reasons outlined by the Standish Group include incomplete requirements, inadequate involvement of end users, the wrong amount of staff with the wrong skills, scope creep, and lack of alignment with strategic objectives of the company.

A project management PCA. This scenario seems ripe for a PCA. Information is stored in a variety of systems distributed throughout the enterprise. Different end-user project management tools hold the project data. Enterprise applications, potentially several of each, perhaps even on different platforms, contain key data about employees, budgets, and products. The current process is manual. If a unified central repository that pulled all project details together, creating a portfolio view, was added, all parties would benefit.

The most obvious vision of a PCA for project management would start with the project managers. The end-user tools used to contain project management schedules would be harvested by the PCA and stored in a central repository. The managers would still use Microsoft Project or other tools to keep track of projects, but they would hit a synchronize button or check in the file, allowing the information to be extracted and stored centrally.

The central repository would also house relevant information from all of the enterprise applications. The repository would create a unified view of status reports, project metrics, employees and their skills, budgets, product specifications, customers, market research, and supplier information. Some of this information would be unique to the PCA, but most of it would be collected from existing systems. Unstructured information like documents, threaded discussions, and selected emails would also be stored in the repository.

The PCA would then provide a variety of functions for the rest of the interested parties. A portfolio dashboard of all projects and their status would be created for senior managers. This dashboard would show the status of projects and enable senior managers to drill down to detailed information about specific projects. The dashboard could be used to monitor and model the capacity of the organization, to confirm that the portfolio of projects reflects strategic priorities, and to assess the impact of changing strategy on the current portfolio of projects. The project managers would have a view of all available employees or resources that were committed, available, or soon to be available. Other participants such as employees, managers, and suppliers would have a view of the information they needed about the projects in which they were involved.

This unified view can then lead to various sorts of action. Executives can constantly review how projects are consuming resources and shift money and people to projects with the most strategic value. Project managers can do the same thing on a project level, moving resources around to get the most of out of the team. If a crucial project step is in danger of failing, then more people can be allocated. Project managers can snap up employees with appropriate skills as they roll off projects that have completed, and employees can request projects that play to their strengths or help them achieve their career development goals.

Product Definition

In Chapter 1 we examined the area of product definition as an example PCA. We will expand on that preliminary discussion and uncover a bit more detail to illuminate how PCAs can make a difference in the modern IT infrastructure.

Scenario: Big money riding on a fragmented product definition process

Product definition involves taking every possible idea for new products or product improvements and categorizing them, evaluating how to produce them, studying their market potential, and then coming up with a short list of promising ideas that match a company's strategy. Those ideas or product concepts contain the seeds of success or failure. Industries like consumer products, high technology, chemicals, financial services, and defense will spend billions of dollars based on the results of the product definition process.

Yet despite the gravity of mistakes in the product definition process, current standard practices leave significant room for improvement. As we explained in Chapter 1, the current product definition process suffers because markets and potential technical problems are not sufficiently analyzed, collection of ideas and customer requirements is not systematic or encouraged properly, ideas are not evaluated by multidisciplinary teams, and product definition information is not stored in a consistent format that allows all of the ideas to be viewed as a comprehensive portfolio. This situation leads to inadequate sharing and reuse of information, and poor timing of product launches and allocation of product development resources.

In this discussion we will examine further the processes that companies have used to address deficiencies in product definition, specifically Dr. Robert Cooper's Stage-Gate™ process.* The Stage-Gate™ process divides product definition into three stages. Before each stage, a gate represents a set of criteria that the product concept must meet in order to be considered for the next stage of analysis. In this way, any idea that pushes through all the gates must meet a rigorous and consistent set of criteria based on a thorough analysis.

The process starts with Gate 1, a gentle preliminary screening by a cross-functional, multi-disciplinary team analyzes the idea from several points of view. Gate 1 has a handful of must-meet and should-meet criteria related to strategic alignment, project feasibility, market potential, and fit with company policies. Ideas that survive this initial screening move into Stage 1, which is called scoping. The scoping stage starts with a preliminary "quick and dirty" assessment of the marketplace, customer requirements, and potential for market acceptance. The scoping phase also includes a preliminary technical assessment that entails in-house approval of the proposed product, assessment of development and manufacturing methods, operations, scheduling, costs, and technical, legal, and regulatory risks and roadblocks.

Following Stage 1, passage through Gate 2 requires a deeper look into the Gate 1 in which analysis along with a search for potential legal, technical, and regulatory obstacles. Financial returns at this stage are simple calculations as opposed to complex models.

Stage 2 builds the business case. It includes detailed market investigation and market research studies. Customer surveys, competitive studies, concept tests, positioning studies, financial analysis, risk assessment, and qualitative business assessment may all be performed. Stage 2 also includes a detailed technical assessment involving preliminary design or lab work that stops short of a full-blown

---

* Cooper's illustration of this process can be found at *http://www.stage-gate.com/Stage-Gate/process.htm*.

prototype. Manufacturing analysis is generally included in Stage 2, investigating manufacturability, source of supply, costs to manufacture, and investment required. If appropriate, detailed legal, patent, and regulatory assessment work is undertaken to remove risks and map out the required actions.

With the business case built in Stage 2, Gate 3 is the last point at which the project can be killed before heavy spending. Gate 3 takes the market-driven product definition developed up to this point and seeks confirmation on the development, operations and marketing plans. Gate 3 scrutinizes the product according to industry-specific functional criteria, platform standards, and interdependencies with the other products or components.

Stage 3 is development of the product prototype, which is where the product definition process ends and the product development process begins, which is generally the purview of the PLM system.

While the Stage-Gate™ process ensures that a thorough analysis is performed, in practice it is frequently not tuned to the special needs of different sorts of product ideas. Should a new product run through the same process as a minor improvement to a stable product? Rigidly applying the same process in all cases can cause delays in bringing products to market or frustrate executives, who may then side-step the process entirely.

A product definition PCA. In Chapter 1 we described a vision for a product definition PCA that employed a single repository of structured and unstructured information for all product definition information across all divisions, avoided duplication of development effort across divisions, assembled information from existing financial and PLM systems, distributed information back to existing systems when appropriate, allowed easy roll-up of information into a portfolio view, encouraged multi-disciplinary collaboration early in the process, including rewarding suggestions that proved valuable, and collected all relevant information for use later in the process or by future product definition teams. We pointed out that such a PCA would allow for a more flexible implementation of structured processes like Stage-Gate™ and would also provide better alignment of product development efforts with corporate strategy and market trends.

Two additional implications of such a PCA merit expanded discussion. First, easy roll-up of product ideas across an enterprise should not only avoid duplicated effort but should also promote better coordination of product definition across the company. Groups that are working on similar ideas will become aware of each other much sooner in the process because of the PCA, research efforts can be shared, and questions and analysis can come from more people with different perspectives. Second, a product definition PCA should promote a circular flow of information between the product definition analysis of the market and technical aspects of the product and the information collected during product prototyping and development. There is no reason that the product definition analysis should stop once prototyping and development has begun. Reactions to the prototype, technical information gained about manufacturing, engineering change requests, and customer research should all be folded back into the product definition analysis, the results of which should continue to guide the product development effort. In this way a virtuous circle of information should be created to improve the end-to-end product development process.

Mergers and Acquisitions

Merging with or acquiring another company is a dramatic move fraught with risk. About half of all M&A transactions fail to increase shareholder value. The reasons deals fall short of expectations are extremely complex and varied, and we will not attempt to catalog every possible cause of failure. This discussion focuses on the process of preparing for and executing an M&A transaction, a process that is ad hoc, supported only sporadically by technology, and ripe for improvement.

Scenario: Disconnected M&A processes with no automation and little shared information

The M&A process has two distinct phases:

1. Identifying candidates, then negotiating and completing the transaction.

2. Executing on the completed transaction by combining the two companies.

One of the major sources of problems in optimizing this process is that the participants in the first phase are highly disconnected from the participants in the second phase.

Identifying targets for an M&A transaction is the domain of corporate development staff, investment bankers, lawyers, and consultants. The senior management team uses the skills and experience of this group to find companies to acquire and then makes the transaction happen. The process involves market research, evaluating the products and competitive position of various companies, and then exhaustive due diligence when a target has been chosen and the transaction is in its final stages.

When the completion of the transaction becomes likely, the managers who will run the merged businesses then meet and plan how the two companies will combine and coordinate their operations. The actions taken at this phase determine whether the merger meets its expectations.

The biggest disconnect in the current processes is that the work done by the team that finds targets and completes the deal rarely is made available to the team who executes the plan to combine the companies. The expectations for how much money will be saved, how operations will be combined, what new cross-selling opportunities will be created, how procurement can be consolidated, what supplier relationships can be improved, and many other such details are collected but not communicated to the execution team. The team doing due diligence may have ideas for executing the merger, but typically they are not conveyed to the execution team. Documents that provide valuable information about how the target company operates are not shared.

The composition of the two teams is different as well. Third parties like lawyers, investment bankers, and consultants comprise most of the team that selects targets and completes the transactions. The permanent members of this team come from the corporate office and tend to have higher turnover. This means that there is generally little institutional memory about the first phase of a merger.

The execution team then starts with a general idea of what the merger is about and seeks to combine the two companies to get the savings or increased revenue outlined by senior management. Even if the execution team had access to the material from the team that completed the transaction, its job would still be daunting. This team faces issues like identifying the same customer in each company's systems; aligning overlapping sales forces; assigning ownership of accounts; communicating with employees, customers, and partners of both companies; and combining procurement, HR, finance, and all other sorts of administrative functions. The teams that execute the merger may change from transaction to transaction and little learning is passed on about how to manage difficult problems.

On top of the disconnects and lack of learning from transaction to transaction, the methods used to manage transactions are generally heavily dependent on end-user tools like spreadsheets, documents, and presentations. These documents are created and distributed through email and shared directories, making finding an authoritative document or rolling up information a difficult process. Process and project management are generally ad hoc, and project managers determine what has been done and what must be done next.

Speed of execution is important. The slower a merger is executed, the more time competitors have to exploit uncertainty and the longer the wait to reap the benefit of any savings or synergies.

A PCA for M&A. A PCA to support the M&A process would be aimed at creating a consciousness of later stages and future transactions to encourage sharing of information across stages. Such a PCA would also create of an institutional memory of how to evaluate and execute such transactions.

The key element of an M&A PCA would be a repository of information that was tightly linked to each stage of the process. The documents created and collected during evaluation of potential targets would be available after a transaction was completed during the execution phase. During phases like due diligence, interesting documents, suggestions, or ideas could be forwarded to the repositories associated with future process steps of the execution phase. The detailed analysis that supported estimates for savings would also be forwarded to the team responsible for producing the savings. This sort of information transfer would provide more information to the execution team and increase accountability for meeting the merger's expectations. Such a structure would also apply the expertise of the third-party consultants to pre- and post-transaction issues.

Processes in all phases of the M&A process could be defined and appropriately automated. Some of the M&A processes are not tightly structured, and collaborative technology could be used to help manage communication and ad hoc activity. More defined processes like playbooks for preparing communication to the interested parties, aligning sales forces, and consolidating procurement could be partially automated at first and improved with experience.

Much of the information could still reside in spreadsheets, documents, and presentations, but, with a centralized repository, searching for and finding such information would be much easier. Key information such as budgets, sales forecasts, and financial models could be extracted from existing systems and used as the starting point for analyzing different combinations of companies. Collaboration and sharing of documents would be supported and the results of collaboration would be recorded.

If such a PCA avoided just a few large mistakes and accelerated the M&A process by making information easier to find and allowing quick identification and removal of bottlenecks, the return on investment would be immediate and substantial. If the transfer of information from previous process steps to later ones saved time and accelerated the completion of various post-transaction processes, even more benefit would accrue from the PCA.

Employee Productivity

In most organizations, employees are frustrated that too much of their jobs are taken up with administrative tasks that distract from their core mission. Many of these tasks are unavoidable, mundane, and inefficient. Automating a small but frequently repeated task can yield significant benefits. Other tasks are important but difficult to perform. People can live without accomplishing them, but in doing so, they sometimes sacrifice the right information or assistance.

One aspect of the points of pain in this area is that they are not show stoppers. The work that is saved by an individual task is not earth-shattering and the new capabilities that may be offered are not revolutionary. The mission of this sort of PCA is to provide a little help in a lot of areas that adds up to a large return on investment, as well as greater employee satisfaction.

Scenario: Mundane tasks add up to sizable burdens and inefficiencies

In companies large and small, it is difficult to find information. In most large companies, the number of documents available for sharing runs into the millions. Information is categorized by who produced it or by the organization chart, not by the topic to which it is related. Applying a simple search engine to this problem results in search results that stretch into the thousands of documents. When knowledge is shared through techniques like email, frequently the information is broadcast widely rather than targeted to those to whom it applies. Each unnecessary email becomes part of information overload and at minimum wastes time.

Another productivity problem relates to changing objectives. After the annual budget is passed, companies typically issue new objectives. The major objectives are distributed across departments and broken down into goals and tasks that are carried out by the staff of the company. Three questions describe the problem with this process: Are the goals feasible and how can information flow to management if they are not? How can we accomplish the goals with our current team? If new skills are required, how will they be acquired? When these questions are asked, there is frequently no forum where they can be discussed and no system to support finding the right skills in the company or arranging to acquire them from without.

Administrative processes such as making vacation requests, ordering paper, and provisioning computers can be time-consuming and error-prone if not automated. Navigating processes that cross functional boundaries like arranging for hiring a new employee can require repeated phone calls to push the process from one department to another in the organization.

A PCA for employee productivity. A PCA for employee productivity could sit on top of existing applications and bring together information from existing systems to help with a variety of lightweight tasks.

One vision for such a PCA is a knowledge network consisting of a master document index, a database of available expertise in the company, survey and polling tools to collect various kinds of feedback, and targeted email groups to ease information overload. An extension of this idea would be an employee interaction center that allows employees to find answers to questions from a knowledge base or to submit questions to a central hub, which can provide automated responses if possible or at least monitor what is being asked and prepare answers that can be published to the knowledge base so employees can find them without assistance. With such a system, resources can be focused on getting a complete answer to the most frequent requests.

The ability of PCAs to span existing systems could be used to automate cross-functional processes such as hiring. Telephone calls that move the process from one step to another could be replaced by a flexible, automated process with email notifications and reminders to keep the process going and move necessary information from one system to another.

All of this functionality is simple and straightforward and would add up to significant savings and happier, more productive employees if applied methodically.

All of the PCAs discussed in this chapter were hypothetical. In the next chapter, we examine PCAs that have been implemented in the areas of project management and product definition.

7

xApps Examples

xApps are SAP's fulfillment of the PCA vision. xApps are products focused on solving specific business problems that build on existing systems to provide new functionality, automating cross-functional processes.

xApps are built on SAP NetWeaver, the company's fulfillment of the ESA vision. SAP NetWeaver is an ESA platform that makes the heterogeneous set of enterprise applications and platform systems into a unified and homogenous foundation on which xApps can be built.

SAP has devoted much effort to explaining its vision for xApps in general, for each particular xApp, and for NetWeaver. Here we summarize the arguments SAP makes in favor of xApps and its ability to produce them. We will also take a tour of xApps that are coming to market and explain how they fulfill and in some ways extend the PCA vision we have discussed in this book.

The xApps Vision

The road to xApps starts with the five C's, which explain what xApps do for a living. According to the five C's, xApps are:

- Cross-functional
- Composite
- Collaborative

- Content-driven

- Closing the loop

"Cross-functional" and "composite" are clear to us now. xApps, like all PCAs, expand the possibilities of automation across functional boundaries by using existing systems. "Collaborative" means that xApps enable ad hoc collaboration, incorporating unstructured information such as email and web conferencing that typically happen outside the scope of process definition. "Content-driven" means that xApps connect knowledge and structured information to a specific business process. "Closing the loop" means that xApps manage the reaction of an organization to an event by assembling the right information, putting it in context, and then providing a framework for orderly collaboration and action that results in a response with maximum business value.

Continuous business innovation is a key part of SAP's vision for how xApps will transform the enterprise. xApps enable a process in which organizations propose a strategy, plan for execution, implement, and then monitor the effects until the cycle begins again with a refinement or adjustment of the strategy. As they appear in the marketplace, xApps will be extended with proprietary technology that will enable rapid implementation of PCAs that support continuous business innovation.

SAP's Strengths in Building xApps

In extending its portfolio of products to xApps and NetWeaver, SAP believes it will be able to build on the adoption of its R/3 ERP system and mySAP enterprise applications and extend its products further into the IT infrastructure. The company points to historical strengths in key areas, strengths that are required for a full implementation of the PCA vision.

SAP's first advantage as an xApp provider is a deep and broad knowledge of the market. The company has more than 20,000 clients in 22 industries across the globe. Implementing SAP technology is not a casual undertaking but requires a deep understanding of a client's business in order to achieve success. The knowledge and relationships gained in thousands of customer engagements gives SAP the ability to define xApps in exactly the right way to provide compelling benefits.

The second advantage SAP has as an xApp provider is 30 years of experience in executing the processes required to define, implement, improve, and support software products. Product requirements flow through product marketing, management, and definition processes, are vetted by influence counsels and interviews with customers, and are tested in pilot implementations. An 8,000-member world-class development team creates the products, which are then supported by a global sales and support team.

SAP's large market share yields a third risk-reducing advantage for creating xApps and the NetWeaver platform. In order to create an ESA platform, a company must develop the framework for unifying all the existing applications, and then create adapters that bring the functionality of those applications and platforms into the ESA platform. SAP has an advantage in providing adapters for its own applications and can amortize the cost of creating adapters for competitors' applications, as well as the entire SAP NetWeaver platform, across the largest installed base in the industry. SAP hopes that it will be the first company with a comprehensive ESA platform across enterprise applications from all vendors, which will make SAP NetWeaver the most attractive way to provide a unified cross-vendor infrastructure. In addition, SAP is likely to be the only vendor offering an open ESA platform that supports both Microsoft's .NET and IBM's WebSphere. SAP believes that compelling xApps that rapidly provide business value will provide the motivation for adopting this infrastructure.

xRPM: An xApp for Project Portfolio Management

The "RPM" in xRPM stands for resource and program management. xRPM provides real-time views into the entire portfolio of projects so management can make informed decisions that synchronize projects with strategy. The system allows optimal deployment of resources to projects, automatic capture of expertise descriptions, and integration of skills with project tasks.

With xRPM in place, companies can identify crucial projects and allocate staff and resources to accelerate the execution and reduce the risk of failure on those projects. In a pharmaceutical company with hundreds of research and development projects, xRPM should reduce the time to market for promising drugs. In large IT organizations, xRPM should reduce the risk of failed projects.

Like all PCAs, xRPM works by adding a layer of processes, services, and objects on top of existing HR, project management, and project costing systems. A simplified architecture looks like Figure 7-1.

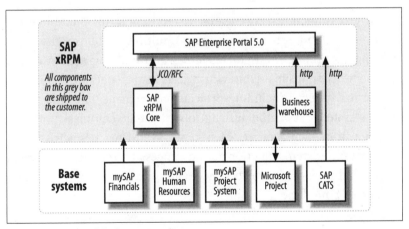

Figure 7-1. A simplified xRPM architecture

From the user's point of view, xRPM is split into four areas:

Project Portfolio Management

Provides visibility across all projects as well as tools for capacity planning based on operational budget, schedule, cost, and staffing information.

Resource Management

Helps managers search for staff, helps employees search for projects, and provides workflows for assignment of resources to projects and impact analysis.

Project Execution

Supports project proposal management, project creation, task assignment, status reporting, and time reporting.

Expertise Management

Allows unique skills discovery and maintenance, and an integrated skills catalog.

Here's how xRPM delivers on the value proposition outlined in the five C's:

Cross-functional

By cross-functional we mean a system that spans multiple organizational boundaries and spans multiple systems. The challenge is finding the right resources and building a team across the functional silos. Strategic projects rely on data from multiple source systems, including project management, HR, and financial systems. It can be difficult to weave together mission-critical data from such a variety of systems. xRPM reaches across organization, system, functional, and process boundaries to create a unified view of all projects and to allow automation of processes across boundaries.

Composite

Organizations should aspire to extend the value and useful life of legacy systems. In project-based organizations, the challenge is how to squeeze new efficiencies out of entrenched project management, HR, and financial systems. To meet this challenge, xRPM provides end-to-end integration that begins with end-user project management systems and extends through enterprise applications.

Collaborative

Projects are inherently team-based endeavors. While it is imperative that intra- and inter-team collaboration be effective, collaboration is often a struggle due to geographic dispersion (time zone differences, no collocation, etc.), a lack of project management standards across the organization, and ineffective tools for communication and collaboration. Today's tools—email, phone, web conferences—are not integrated with systems of record. xRPM solves this problem with integrated collaborative functionality.

Content-driven

Effective project planning and project execution depend on accurate information. Accurate information is difficult to come by, though, in an environment that relies on heterogeneous, disjointed systems with nonstandard reporting practices. Without quality information, it's very difficult to plan new ventures or keep existing projects on track. Critical to every proposal and every running project is information on scope, schedule, and resources. A change in any one of these elements affects the other two. If resources leave the project, schedule and scope may be affected. To succeed, a project-based organization needs real-time insight into mission-critical content. xRPM provides exactly this.

Closing the loop

In a utopian world, we know that corporate strategy should drive the project portfolio. Conversely, project results should help refine corporate strategy. However, in practice, this rarely happens because projects cannot be viewed as a complete portfolio and projects results are not visible. xRPM offers both of these features. Managers are armed with a high-level view of projects as well as the ability to drill down into individual projects, providing a foundation for decisive action to manage projects and influence strategy.

xRPM Case Study

PharmaLarge is a fictional global pharmaceutical giant. Like all major companies in the industry, PharmaLarge labors under intense financial pressure due to high overhead and lengthy development timeframes. For example, the cost of producing a new drug, which can take up to 12 years to develop, can range from $250 million to $500 million, a typical spectrum according to the Pharmaceutical Research and Manufacturers Association of America. Of those drugs that make it to market, however, only a small fraction will reach annual sales over $350 million. To recoup this investment and maintain a 10% growth rate, PharmaLarge must launch about four successful products annually—a far cry from the current average of one per year.

Under these conditions, PharmaLarge cannot afford wasted opportunities, duplicated efforts, or stalled projects. However, such problems have been mounting. The company's homegrown, 15-year-old method for tracking project performance is failing. This system relied primarily on spreadsheets maintained in the company intranet and updated on an irregular basis. It provided no communication between disparate data systems such as financials and HR and team collaboration was bogged down by confusion, delays, and a lack of prioritization.

PharmaLarge implemented SAP xRPM to wholly revise portfolio management, boost project performance, and speed time to market for new drugs. Figure 7-2 shows its project portfolio.

PharmaLarge executives are now better equipped to make portfolio and project investment decisions because SAP xRPM summarizes mission-critical data from various backend systems. Before xRPM, PharmaLarge spent three days per month culling data from disjointed HR, financial, and project management applications. Today, the company taps into the same information in real-time.

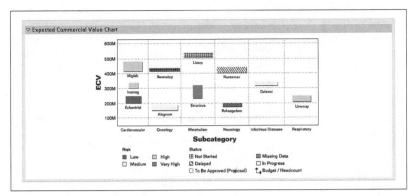

Figure 7-2. PharmaLarge project portfolio

Project leaders now consult their project portfolio dashboard (see Figure 7-3) for current data on every project. Project managers can keep tabs on project risk, schedule performance, and staffing levels.

Subcategory	Projects	Priority	Budget	Schedule	Risk	Staffing	Phase
Cardiovascular	3		△	◉	▣	▣	
	Migleb	Medium	△	▣	△	▣	Not started
	Eubantrist	High	◉	◉	▣	▣	Pre-clinical
	Inameg	High	◉	◉	▣	△	Phase II
Oncology	2		△	△	△	◉	
	Alognom	Medium	△	▣	△	◉	Phase I
	Xwanatop	High	◉	◉	◉	◉	Discovery
Metabolism	2		◉	◉	△	◉	
Neurology	2		▣	▣	△	◉	
Infectious Diseases	1		◉	△	△	△	
Respiratory	1		◉	△	▣	◉	

Figure 7-3. PharmaLarge project portfolio dashboard

PharmaLarge's R&D manager, Joe Wallace, uses this dashboard to quickly determine which projects are on schedule, over budget, or have staffing shortfalls. From the dashboard, he drilled down into a more detailed analysis of how the new drug, Exce, was doing. He noticed that development had fallen behind schedule (see Figure 7-4).

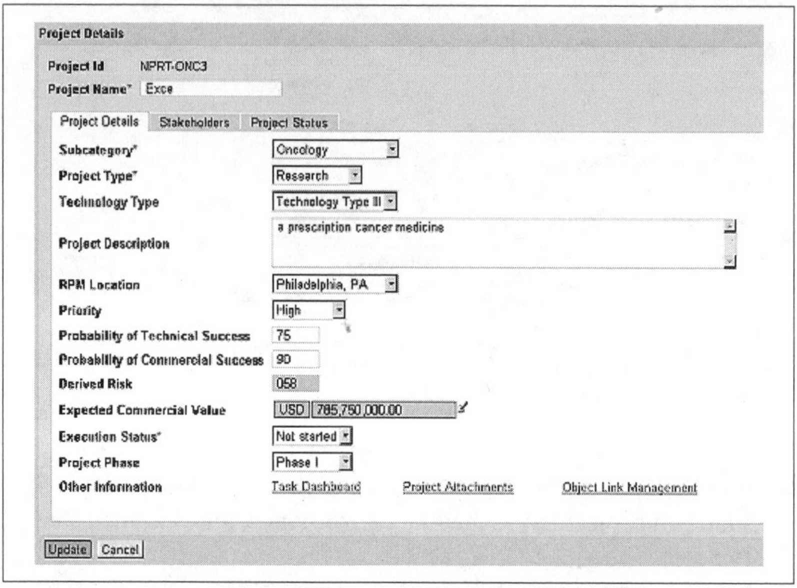

Figure 7-4. Using the project portfolio dashboard

Joe surveyed detailed reports on the project team and saw that the group was in need of two more oncology specialists. Such an addition would entail a significant budgetary increase for the project. Drilling more deeply into the Exce detailed report, Joe saw the information indicating that Exce was rated low risk with a high probability of commercial success. Realizing the value of getting the project back on track, he allocated more funding and notified the Exce project manager.

Jane Phillips, the Exce project manager, received this information and quickly used SAP xRPM to review eligible oncology specialists within the company. She found only one specialist in her geographic region who was available to take on a new project and whose skill set was a close match. Jane then contacted the HR specialist in her area to assist with the search and hire of an outside oncology expert. The HR specialist was automatically supplied with the detailed skill requirements.

Meanwhile, the internal oncology specialist, Alex Smart, received Jane's query, the Exce detailed report, the skills required, and the team information. He compared the skills and knowledge required

to his and to his self-designed career map. Alex saw that the pr
would considerably advance his knowledge in an area of cu
research in which he was highly interested. Noting that his current
project would conclude in three weeks, he accepted the offer.

Using SAP xRPM, the project team was able to swiftly resolve a
project delay by identifying the source of the problem and taking
appropriate action. Exce went on to complete the discovery phase on
time and then entered phase I clinical trials.

In short, PharmaLarge uses SAP xRPM to expertly forecast and
match skills requirements, track project performance against budget
and value, and cull and sort performance data by topical area and on
a project-by-project basis. As a result, the company expects to save
millions of dollars in recovered productivity, enhanced product quality, and swifter time to market.

xPD: An xApp for Product Definition

xPD is focused on automating and improving the product definition
process. Product definition is the first phase of the larger process of
developing products in which ideas and customer requirements are
analyzed and converted into product concepts, which are in turn
evaluated for their market potential and technical characteristics.
Mature product concepts then enter the design phase, which starts
with prototyping and ends with production.

Making the right decisions about which products to develop is one
of the key processes in many different types of companies. Much
analysis and study has gone into making this process as effective as
possible. The most popular methodology for product development is
the Stage-Gate process designed by Robert Cooper, which describes
several stages of product development and several gates, or sets of
criteria, that a product concept must pass through in order to qualify for development.

Like many strategic processes, current product definition processes are frequently managed through end-user tools like spreadsheets and word processors. The documents created in such a manner are then shared using email or file servers. The processes for product definition are defined at a general level, like the Stage-Gate process, but are not supported through any significant automation. This reduces the visibility into the product definition process and makes a comprehensive portfolio view of all ideas and requirements difficult to come by. If a roll-up of ideas is done, it is generally time-consuming, manual, and quickly goes stale. The same manual methods are used for extracting information from existing systems like CRM for customer requirements, ERP for financial and budget information, and PLM for product specifications.

xPD provides a general framework in which a process like Stage-Gate can be automated and fine-tuned to different needs and products. xPD can also create a unified repository for structured and unstructured information and a collaborative environment that brings the right people into each stage of the process.

xPD serves users in three important roles: business developers, program managers, and design managers. Each has different needs and desires. Business developers find product concepts that capitalize on new market trends and manage customer requirements. They ask questions like: How successful are my products? What are the market trends and what innovative product concepts will capitalize on them? How can ideas and customer requirements be managed and categorized?

Program managers take the concepts defined by business developers and determine which ones are the most promising. They flesh out product concepts by describing detailed product requirements, determining market size, and assembling the right multi-disciplinary team to make sure that every problem or opportunity is identified. Business developers then guide proposed products through the approval process.

Design managers assemble teams from inside and outside the company to carry out detailed product design and determine the final production cost for the product. Design managers must ensure that any changes during design result in a product that meets the business requirements.

The xPD application helps these individuals through process automation, integration with existing systems, idea and product portfolio analysis, and collaborative functionality. Business developers are provided with functions that perform product portfolio analysis, surveys about new product features, analysis of ideas, and automatic classification and consolidation of ideas into categories.

Program managers can use functions that aid in defining and prioritizing product concepts, collaborative services for chats and online voting, and project management and process templates for building business cases and carrying out feasibility studies. Special support is provided for extracting and organizing information on market trends.

Design managers can use project and process templates and collaborative technology for keeping track of discussion threads related to the design and testing results for the prototypes. Changes in requirements and design are recorded and approved starting from the initial design, through the construction of the prototype and through every phase of development and manufacturing, including engineering changes. Information from existing systems is vital to the design phase so integration with PLM, ERP, and other systems is provided.

xPD helps companies develop the right products in the most efficient manner, avoiding foreseeable mistakes, decreasing time to market, and reducing development costs and failure rates by applying the right skills and expertise early in the process. Efforts across divisions are better coordinated because of xPD's comprehensive portfolio view. The investment in existing systems is leveraged because information from those systems is brought into the process right when it is needed.

xPD implements the 5 C's as follows:

Cross-functional

xPD drives complete idea-design processes across multiple existing applications and company boundaries. It facilitates bringing in experts from the legal department, quality management, manufacturing, finance, and market research early enough in the process to save significant time and money.

Composite

Information from existing enterprise applications like PLM, ERP, and CRM as well as legacy applications is integrated into xPD and reused, amplifying the value of these systems.

Collaborative

xPD enables and automates collaboration across all internal and external groups. The results of the collaboration are saved to record why a product concept was approved or rejected.

Content-driven

xPD creates a comprehensive portfolio view with all required information to replace the fragmentation of information that results when end-user applications are the primary repositories.

Closing the loop

The customized environment of xPD gives each person the ability to review appropriate information and then take action. The results of those actions show up in xPD and also in the enterprise applications and legacy systems so that the effect of the actions resonate throughout the enterprise.

xPD Case Study

Novellus is a fictional global high-tech giant. Like all major companies in the industry, Novellus labors under intense financial pressure due to short product lifecycles, high product development costs, and critical time-to-market delays. The company's largest problem is a significantly higher new product failure rate than that of its competitors. Despite the fact that the company spends about 11 percent of revenues in R&D, Novellus derives only 25.2 percent of sales from new products while others in the industry gain approximately half their revenues from that source. Furthermore, the great majority of Novellus's products never make it to market. Those that do have a failure rate somewhere between 30 to 40 percent.

Under these conditions, Novellus cannot afford wasted opportunities, ineffective products, and an inefficient development process. However, it is clear to the company's management that too much R&D spending is wasted. Novellus spends about 15 percent of its product innovation resources on exploratory research, idea generation, and initial attempts to qualify the idea. Most of this goes to failed products. The lack of company-wide structured processes for evaluating product concepts entering development means that most concepts are rapidly—and prematurely—pushed into development. Sometimes products are recognized as doomed for failure, but only after considerable resources have already been spent. These critical market and evaluative processes did not have proper IT support. The idea and requirements generation system relied primarily on spreadsheets maintained on the company intranet and updated on an irregular basis. Requirement changes were often inaccurate, and sometimes were lost altogether. Collaboration across organizational, geographic, and functional lines was limited to desktop programs such as Outlook and PowerPoint. Niche solutions purchased during the dotcom heyday were on the shelf due to integration or support issues. Communication between disparate data systems (even within different PLM solutions) was not possible, and process flow was bogged down by confusion, delays, and a lack of prioritization.

Novellus implemented SAP xPD to wholly revise product portfolio management and enhance idea and concept generation; the company hoped to develop highly effective new products.

Novellus executives are now better equipped to make product portfolio analyses and new development investment decisions because SAP xPD summarizes mission-critical data from various backend systems. Before xPD, executives could see only high-level portfolio information on PowerPoint presentations. They lacked the level of detail required to approve new concepts or kill nonstrategic projects.

Business developers now leverage their portfolio dashboards to assess and consolidate a promising pool of ideas and customer requirements from different sources. For example, Novellus Business Developer Pat Smith uses this dashboard to quickly determine the status of his product lines in the marketplace. From the dashboard, he can drill down into a detailed analysis of the latest quarterly information on the company's PDA product line charted against market information (provided as a web service from AC Nielsen). He notices that the company's market share has been eroding. Figure 7-5 provides an overview of product line enhancement ideas.

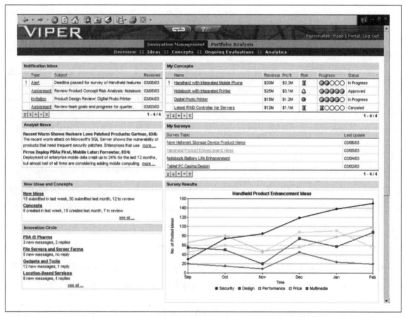

Figure 7-5. Product concept ideas summarized by product line

In response, Pat surveys detailed reports on ideas, customer requirements, and market research, including competitive information. He realizes that recent customer feedback and industry reports mention unreliable security features in the PDA products as a competitive disadvantage. He drills down into all security-related ideas in the xPD repository (with all respective documentation) and finds a collection of promising submissions recommended by Novellus's security software suppliers. He also realizes, to his astonishment, that feedback from manufacturing about platform compatibility issues with existing security products have not been addressed properly. Equipped with this vital information, Pat submitted a product upgrade request to the initial idea approval committee, which gave the project high priority (see Figure 7-6).

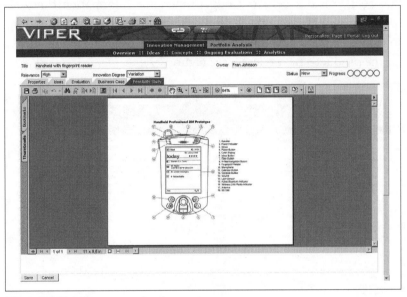

Figure 7-6. Product concept detail

Deborah Thatcher, a PDA program manager, received this information and quickly used SAP xPD to review all available PDA and security specialists within the company. She also scanned for all relevant cross-functional employees, especially the manufacturing engineers with the initial requirement change request, who would be required to undertake the necessary market and technical assessments. Each of these team members was automatically sent workflow alerts with all the relevant background documentation, project needs, and deadlines as well as suggested templates and reference analytic tools. Other program managers were also alerted of the executive decision of the prioritization of this project for justification of resource allocations.

Meanwhile, individual team members were able to conduct their necessary tasks in much shorter time than usual because they had all the necessary information, access to all the right people, and were relieved of other low-priority projects. For example, the in-house security specialist, John Travers, received Deborah's task alert with all documentation. He was also informed of collaborative tasks to be undertaken with the security software suppliers and design and man-ufacturing engineers to address all product interdependencies. For that he also had access to product functional information from the underlying PLM system. With such effective cross-collaboration, John was able to specify customer-centric functional requirements for the product's next release (see Figure 7-7).

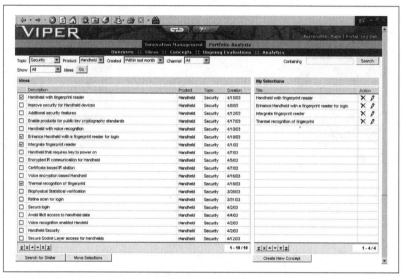

Figure 7-7. Security-related product ideas

With the evaluative reports consolidated from the entire team and the relevant product portfolio information, the approval committee was able to make a swift decision on committing more development resources for prototype design and development. This decision trig-gered a workflow whereby the design manager, Peter Landsey, was alerted with all the functional requirements information. All this was also automatically (based on executive role decisions) written back into the PLM system where Peter utilized the CAD modules for the prototype design. He was also able to address minor functional requirement changes needed due to some platform incompatibility issues. This information was resent to the product definition team for further review, ensuring the critical bi-directional sharing of information between the different stages.

Using SAP xPD, the product development team was able to swiftly address product deficiencies by identifying the market needs and taking appropriate action. The enhanced PDA was released in the market a month ahead of schedule to great customer feedback and favorable market reviews.

In short, Novellus uses SAP xPD to make intelligent portfolio analyses and product development choices based on real customer needs, the best ideas, and in-depth analyses. As a result, the company expects to save money and increase revenue with market-driven winning products, recovered productivity, and swifter time to market.

xPD provides what most xApps provide: packaged integration, collaborative functionality, and flexible process automation focused on a vital process. Given the stakes, if xPD provides just a small improvement, the return on investment should be huge.

The Growing xApps Portfolio

The xApps mentioned in this chapter are the beginning of a stream of products from SAP and other vendors. xApps are on the way for mergers and acquisitions, employee productivity, automation of plant maintenance, improving the supply chain in the petroleum industry, and a variety of other areas. ISVs and systems integrators are planning to offer xApps, and companies are signing up to do custom development based on SAP NetWeaver. In the next and final chapter of this book we will examine what the world will look like as xApps and PCAs propagate through the IT infrastructure.

The Future of PCAs

One future for PCAs is a forgotten dream. Many ideas arrive, have their day in the sun, and then drift off into oblivion.

But the notion of PCAs seems a bit more durable. If corporations do not build the next generation of applications on top of their current infrastructure, if they do not make applications more flexible with configurable processes, if they do not expand the scope of automation with finer-grained applications, then what will they do?

Our bet is that PCAs will shape application delivery. The deeper changes to the structure of core enterprise applications and the infrastructure that connects them will evolve as well. The second book in this series covers Enterprise Services Architecture and takes a closer look at how that evolution will take place and the forces at play. While architectural evolution is a long process, the value of PCAs will be felt in the short term, as outlined in Chapter 4.

To build on our discussion from Chapter 5 on the likely effects of PCAs, we asked Yvonne Genovese, Vice President and Research Director at Gartner, Inc., what she thinks the future holds for PCAs and how they will change the landscape for enterprise computing.

"PCAs are the marriage of IT and business," she said. "This is where it happens, at the PCA level."

Because PCAs will be finer-grained and designed to deliver a targeted piece of functionality, Genovese predicts implementation cycles will be faster and the return on investment more rapid. The granularity will also result in the growth in metrics at a more detailed level of all business processes. Current metrics, for example, might record how fast a product was delivered once it was ordered. PCA metrics will show the length of time in each portion of the order: taking the order, determining how to fill it, assembling the components or services needed to satisfy the order, and ultimately delivering it.

Genovese's view is that PCAs are the first significant ripple in a slow-motion tidal wave of change that will eventually engulf the enterprise. Her vision is that PCAs will bring executives what they have always wanted: a unified, real-time view of the enterprise that allows them to take action and direct the operations of the business. The ability of PCAs to assemble information across all enterprise applications and their configurability and agility will accelerate strategic decision-making processes. In essence, if the corporation is a body, then PCAs will help speed up the thinking process and allow management to understand the nature of their business in a way never before possible.

The result of this new information will be, for most companies, a new vision for how business should be shaped, which will involve changes in core operations. Executing on that vision, propagating the agility of PCAs to all aspects of the enterprise, will be a slower process. If crafting the new vision takes a year, then implementing it may take ten. Perhaps this is similar to the difference between thinking up a new weight training program and actually transforming your body over a long period by lifting weights.

How will we know if PCAs are changing the world of IT? Genovese suggests watching for a few key trends and indicators. She predicts that because PCAs are more targeted and contain and implement business processes, IT vendors will increasingly sell only to the business stakeholders instead of to the technologists. She predicts that the executives in corporations will be the biggest initial beneficiaries as PCAs turn the lights on in the enterprise and provide an integrated view of operations at their companies. This spotlight of information will expose inefficiency at all levels: in departmental processes, in vendor performance, in the supply chain, and in strategic processes. She predicts that IT vendors who have a sufficient critical mass of customers or those with a precise focus will have an advantage in satisfying executives with an appetite for PCAs.

If Genovese is right, then there is probably a PCA in the future for most executives focused on gaining significant strategic and operational benefits from information technology. We hope that this book will provide a head start in taking advantage of PCAs, help avoid some significant problems, and ultimately gain for your company all of the benefits that PCAs have to offer.

Index

We'd like to hear your suggestions for improving our indexes. Send email to *index@oreilly.com*.

About the Author

Dan Woods is CTO and Publisher of the Evolved Media Network, a consulting and publishing firm that provides services for technology communications. Dan has a background in technology and journalism. He has a BA in computer science from the University of Michigan. He was CTO of *TheStreet.com* and CapitalThinking, led development at Time Inc.'s Pathfinder, and created applications for *NandO.net*, one of the first newspaper web sites. Dan has an MS from Columbia University's Graduate School of Journalism. He covered banking for three years at *The Record of Hackensack*, was database editor for three years at the *Raleigh News & Observer*, and has written several books on technology topics, in addition to numerous white papers and magazine articles. He lives in New York City with his wife and two children. Dan can be reached at *dwoods@EvolvedMediaNetwork.com*.

Colophon

Our look is the result of reader comments, our own experimentation, and feedback from distribution channels. Distinctive covers complement our distinctive approach to technical topics, breathing personality and life into potentially dry subjects.

Darren Kelly was the production editor, Debra Cameron was the developmental editor, and Jane Ellin was the proofreader for *Packaged Composite Applications*. Claire Cloutier provided quality control. Judy Hoer wrote the index.

Edie Freedman designed the cover of this book. The cover image is an original engraving from the 19th century. Emma Colby produced the cover layout with QuarkXPress 4.1 using Adobe's ITC Garamond font.

David Futato and Edie Freedman designed the interior layout. This book was converted by Andrew Savikas to FrameMaker 5.5.6 with a format conversion tool created by Erik Ray, Jason McIntosh, Neil Walls, and Mike Sierra that uses Perl and XML technologies. The text font is Linotype Birka; the heading font is Adobe Myriad Condensed; and the code font is LucasFont's TheSans Mono Condensed. The illustrations that appear in the book were produced by Robert Romano and Jessamyn Read using Macromedia FreeHand 9 and Adobe Photoshop 6.

About O'Reilly

O'Reilly & Associates is the premier information source for leading-edge computer technologies. The company's books, conferences, and web sites bring to light the knowledge of technology innovators. O'Reilly books, known for the animals on their covers, occupy a treasured place on the shelves of the developers building the next generation of software. O'Reilly conferences and summits bring alpha geeks and forward-thinking business leaders together to shape the revolutionary ideas that spark new industries. From the Internet to XML, open source, .NET, Java, and web services, O'Reilly puts technologies on the map.

For more information, see *www.oreilly.com*

O'REILLY®

To order: *800-998-9938* • *order@oreilly.com* • *www.oreilly.com*
Online editions of most O'Reilly titles are available by subscription at *safari.oreilly.com*
Also available at most retail and online bookstores.